Precision Machining Technology Workbook and Projects Manual

David Lenzi

Lone Star College, University Park, Houston, TX

James Hillwig

Crawford County Career Technical Center, Meadville, PA

DELMAR
CENGAGE Learning™

Australia • Brazil • Japan • Korea • Mexico • Singapore • Spain • United Kingdom • United States

DELMAR
CENGAGE Learning™

**Precision Machining Technology Workbook
and Projects Manual**
David Lenzi, James Hillwig

Vice President, Editorial: Dave Garza
Director of Learning Solutions: Sandy Clark
Executive Editor: David Boelio
Managing Editor: Larry Main
Senior Product Manager: Sharon Chambliss
Editorial Assistant: Jillian Borden
Vice President, Marketing: Jennifer Baker
Marketing Director: Deborah Yarnell
Marketing Manager: Kathryn Hall
Production Director: Wendy Troeger
Production Manager: Mark Bernard
Senior Content Project Manager: Cheri Plasse
Senior Art Director: Joy Kocsis
Technology Project Manager: Christopher Catalina

For product information and technology assistance, contact us at
Cengage Learning Customer & Sales Support, 1-800-354-9706

For permission to use material from this text or product,
submit all requests online at **www.cengage.com/permissions**
Further permissions questions can be emailed to
permissionrequest@cengage.com

Example: Microsoft® is a registered trademark of the Microsoft Corporation.

Library of Congress Control Number: 2010930755

ISBN-13: 978-1-4354-4768-4

ISBN-10: 1-4354-4768-9

Delmar
Executive Woods
5 Maxwell Drive
Clifton Park, NY 12065
USA

Cengage Learning is a leading provider of customized learning solutions with office locations around the globe, including Singapore, the United Kingdom, Australia, Mexico, Brazil, and Japan. Locate your local office at **www.cengage.com/global**

Cengage Learning products are represented in Canada by Nelson Education, Ltd.

To learn more about Delmar, visit **www.cengage.com/delmar**

Purchase any of our products at your local bookstore or at our preferred online store **www.cengagebrain.com**

Notice to the Reader
Publisher does not warrant or guarantee any of the products described herein or perform any independent analysis in connection with any of the product information contained herein. Publisher does not assume, and expressly disclaims, any obligation to obtain and include information other than that provided to it by the manufacturer. The reader is expressly warned to consider and adopt all safety precautions that might be indicated by the activities described herein and to avoid all potential hazards. By following the instructions contained herein, the reader willingly assumes all risks in connection with such instructions. The publisher makes no representations or warranties of any kind, including but not limited to, the warranties of fitness for particular purpose or merchantability, nor are any such representations implied with respect to the material set forth herein, and the publisher takes no responsibility with respect to such material. The publisher shall not be liable for any special, consequential, or exemplary damages resulting, in whole or part, from the readers' use of, or reliance upon, this material.

Printed in the United States of America
2 3 4 5 6 7 15 14 13 12 11

Table of Contents

PROJECTS

Preface

The *Precision Machining Technology Workbook and Projects Manual* is intended to assist the student in the understanding and development of the fundamental and intermediate machining skills needed for career success in a rapidly changing manufacturing environment.

The Workbook section of the manual is designed to reinforce the student's understanding and comprehension of the material presented in each unit of the text. After reading and completing each unit of the text, the student should be prepared to successfully complete each corresponding Workbook unit.

The Projects section of the manual used in conjunction with the text offers a wide range of material to aid in hands-on performance as it relates to NIMS credentialing. At the beginning of each project there is a list of the various NIMS skills practiced, bringing the student a deeper understanding of that skill as each project is completed. The projects are grouped together so that each NIMS skill can be mastered. Upon successful completion of the projects, the student will be sufficiently prepared to take the corresponding NIMS performance test.

About the Authors

David Lenzi, Lone Star College, University Park, Houston, TX

David has 16 years of manufacturing experience and 11 years of post-secondary administration and instruction experience. He holds three degrees: Associate of Applied Science in Tool & Manufacturing Technology, a bachelor's degree in Industrial Technology, and a master's degree in Business. He has completed a tool, die, and mold maker apprenticeship; has designed defense tooling; has taught precision machining in Texas, Illinois, and Missouri; has held a quality engineer position; and currently resides in the Houston, Texas, area, working for Lone Star College as Senior Program Manager in workforce development.

Jim Hillwig, Crawford County Career Technical Center, Meadville, PA

Jim has had many years of teaching experience, both at the secondary and post-secondary levels. He has been working with NIMS since 1998 in developing and implementing the NIMS curriculum for his school and the state of Pennsylvania. He currently holds 10 NIMS Machining Level I certifications and 8 NIMS Machining Level II certifications.

SECTION 1: Introduction to Machining

UNIT 1	Introduction to Machining

Name: _____ Date: _____

Score: _____ Text pages 1–19

Learning Objectives

After completing this unit, you should be able to:

- Define the term *machining*
- Define a machine tool
- Discuss the evolution of machining and machine tools
- Identify the role of machining in society
- Discuss the principles of the basic types of machining processes

Carefully read the unit, and then answer the following questions in the space provided.

Fill in the Blank

1. The _____ probably was the first and the simplest of all machine tools.
2. The average American worker in the manufacturing sector earned over $ _____ annually, including benefits, while the average non-manufacturing worker earned under $ _____ annually.
3. The _____ machine uses abrasive grit that is introduced into a very high-pressure, focused jet of water to perform cutting.
4. _____ are classified as CNC machine tools and can cut very hard materials easily.
5. The most common types of precision grinders used in a machine shop are the _____ and _____.

Short Answer

6. When defining the word "Machining" using a standard dictionary, what two key defining phrases are concluded?

   _____

   _____

7. Choose a common item that you use or see every day. Briefly research how the machining industry is related to that item and report your findings.

   _____

   _____

8. What major advancement in machine tools occurred in the 1970s and what benefits did it provide?

   _____

   _____

9. List four reasons why we need accuracy in machined parts.

_____

_____

10. Briefly describe the role of machining in today's society.

_____

_____

True or False

11. Animals and/or water powered the earliest machine tools.   _____

12. Machining is connected to the manufacturing of durable goods.   _____

13. Some machining operations can produce sizes to ±0.0001 inch or less of the desired size.   _____

14. The U.S. is still the leading manufacturing nation of the world.   _____

15. There are nearly 153.7 million people directly employed in manufacturing jobs in the U.S.   _____

16. A sawing machine is never used for precise machine cutting.   _____

17. Milling machines can be either vertical or horizontal.   _____

18. Electricity can be used to machine material to desired shape.   _____

Identification Matching

19. Match the appropriate inventor, name to their invention.

a. Henry Maudslay	_____	**1.** Creation of the assembly line for mass production
b. John Wilkinson	_____	**2.** Invented machine to cut accurately screw threads
c. John Smeaton	_____	**3.** Produced the first milling machine
d. Eli Whitney	_____	**4.** Developed the Great Wheel lathe
e. Henry Ford	_____	**5.** Developed a water powered boring machine

20. Match the term with the correct machine tool definition.

a. Durable goods	_____	**1.** Parts are produced from tools called "molds"
b. Consumable goods	_____	**2.** High-tech machining operations are used to produce complex parts
c. Plastic goods	_____	**3.** Parts usually made with one type of die—a forming die
d. Medical goods	_____	**4.** Industries rely heavily on machining and machined parts
e. Transportation goods	_____	**5.** Equipment that produces clothing and paper

21. Select the appropriate machine tool process to produce the given product.

a. Lathe	_____	**1.**	Uses grinding wheels for either precision or non-precision machining
b. Milling machine	_____	**2.**	Uses a highly concentrated beam of light
c. Drill press	_____	**3.**	Performs hole-making operations by feeding various types of rotating cutting tools
d. Bandsaw	_____	**4.**	Electricity used to machine a part
e. EDM	_____	**5.**	Used to cut material to rough lengths or to remove large sections of material quickly
f. Laser machining	_____	**6.**	Uses abrasive grit at very high pressure
g. Water jet machining	_____	**7.**	Used to produce cylindrical parts
h. Abrasive machining	_____	**8.**	Uses rotating cutters moved across a part to remove material

Multiple Choice

22. The cord of a bow string drill was used for
   a. holding the cutting tool.
   b. providing power for rotation.
   c. transportation of the drill.
   d. finding center of the hole.

23. In the early 1900s, Henry Ford developed the first
   a. lathe to cut threads.
   b. automobile.
   c. mass production assembly line.
   d. steam power engine.

24. In the 1970s, the machining industry changed forever with the invention of the
   a. punch tape recorder.
   b. numerical control machine tool.
   c. electric power feed.
   d. computer controlled machine tool.

25. Manufacturing plays an important role in which of the following?
   a. Economy of the country
   b. U.S. workforce
   c. Standard of living
   d. All of the above

SECTION 1: Introduction to Machining

UNIT 2	Careers in Machining

Name: _____ Date: _____

Score: _____ Text pages 20–27

Learning Objectives

After completing this unit, you should be able to:

- Identify and discuss careers in the machining industry
- Identify and discuss careers in fields related to machining
- Discuss the job outlook in the machining field
- Understand and explain effective job-seeking skills

Carefully read the unit, and then answer the following questions in the space provided.

Fill in the Blank

1. The _____ forecasts the future of specific job markets.
2. Industrial _____ careers exist to promote and market machines, machine tools, and accessories.
3. The _____ engineer continually studies the production process for improvement engineering.
4. Entry-level CNC employees who are beginning careers in CNC machining, and have little prior knowledge of CNC machining, are called _____.

Short Answer

5. How have occupations in the machining field changed over time?
   _____
   _____

6. Compare and contrast CNC and conventional machinists. How are their jobs similar? How are they different?
   _____
   _____

7. Briefly discuss the similarities and difference between die makers, mold makers, and toolmakers.
   _____
   _____

8. What type or types of machining careers might be of interest to you? Why? Which ones would not interest you? Why not?
   _____
   _____

9. Describe how CAD/CAM has changed machining manufacturing.
   _____
   _____

True and False

10. CNC operators are highly skilled machinist.   _____

11. Machining careers provide advancement opportunities to management and engineering.   _____

12. The term machinist and mechanic are interchangeable.   _____

13. The job outlook for machinist and tool makers has been positive.   _____

14. CNC operators are responsible for the quality of product not the development
of the CNC part program.   _____

Identification Matching

15. Match the career job role to the appropriate job duty.

a. Set technician	_____	**1.** Uses CAD software to draw models or engineering drawings
b. Operator	_____	**2.** Makes components for either plastic or die-cast metal parts
c. Conventional machinist	_____	**3.** Makes complex tools, jigs, and fixtures
d. CNC machinist	_____	**4.** Selects proper cutting tools and devices
e. Programmer	_____	**5.** Makes cutting tools consisting of punches and dies
f. Toolmaker	_____	**6.** Experiences running almost every type of machine tool
g. Die maker	_____	**7.** Responsible for planning, scheduling, and purchasing
h. Mold maker	_____	**8.** Possesses the skills of both the set-up technician and the operator
i. Supervisor	_____	**9.** Places parts in machines and continually runs a set operation
j. Designer	_____	**10.** Writes programs consisting of machine code

Multiple Choice

16. The most experienced employee in a traditional machine shop would have the title of
  a. machine operator.
  b. set-up technician.
  c. tool maker.
  d. CNC machinist.

17. A career field that requires troubleshooting and problem-solving skills as well as understands the function
and repair of mechanical, electrical, electronic, hydraulic, and pneumatic systems is
  a. set-up technician.
  b. field service technician.
  c. industrial salesman.
  d. quality control inspector.

18. A career field title that focuses on the manufacturing process is
  a. mechanical engineer.
  b. quality engineer.
  c. design engineer.
  d. manufacturing engineer.

SECTION 1: Introduction to Machining

UNIT 3 | Workplace Skills

Name: _____ Date: _____

Score: _____ Text pages 28–41

Learning Objectives

After completing this unit, you should be able to:

- Identify and understand personal skills needed for success in the machining field
- Identify and understand technical skills needed for success in the machining field
- Show understanding of training opportunities and methods available to gain skills required for the machining field
- Create a career plan
- Create a resume
- Create a cover letter
- Compile a list of references
- Create a thank-you letter
- Describe a portfolio and its importance
- Use different methods to find job opportunities
- Conduct a practice interview

Carefully read the unit, and then answer the following questions in the space provided.

Fill in the Blank

1. _____ is the ability to make judgments using many pieces of information.
2. A program that contains a certain number of hours of practical training in machining operations during normal working hours is called a _____.
3. Four year degrees are usually granted from a _____.
4. Some employers provide training to employees while they are receiving wages; this is called _____.

Short Answer

5. Write a one-page career plan.
6. List five personal skills.

_____
_____
_____
_____
_____

7. List the three parts of a cover letter.

_____
_____
_____

8. Think about the personal workplace skills discussed in this unit related to you. Which skills do feel you might already possess or are your current strengths? Where do you think you might need to improve?

_____

_____

9. Start collecting items for a career portfolio. Include letters of reference, special awards, and photographs of your work.

10. Briefly explain the difference between a journeyman and apprentice.

_____

_____

11. What is the primary role of NIMS?

_____

_____

True or False

12. Instant gratification and completion is not usually the norm for those in the machining field.   _____

13. Even small lapses in attention can lead to large errors and huge losses of time and money.   _____

14. The mechanical aptitude is a combined eye-hand coordination to control motion.   _____

15. All machining careers require a combination of mental and hands-on skills.   _____

Multiple Choice

16. The combination of mental and physical skills a machinist must have is called
    a. mechanical aptitude.
    b. manual dexterity.
    c. eye-hand coordination.
    d. physical ability.

17. The ability to communicate effectively and to share information among peers is called
    a. persistence.
    b. interpersonal skills.
    c. public speaking.
    d. interpretation.

18. Companies may also need to provide training for specific specialized areas and may send employees off-site or bring trainers on-site to meet that need; this kind of training is called
    a. RJT.
    b. JTO.
    c. OJT.
    d. JOT.

19. An employee that is expected to be able to perform any machining operation required by a company is called a(n)
    a. machine operator.
    b. apprentice.
    c. journeyman.
    d. programmer.

SECTION 2: Measurement, Materials, and Safety

UNIT 1 | Introduction to Safety

Name: _____ Date: _____

Score:_____ Text pages 42–62

Learning Objectives

After completing this unit, you should be able to:

- Define *OSHA* and describe its purpose
- Define *NIOSH* and describe its purpose
- Describe appropriate clothing for a machining environment
- Indentify appropriate PPE used in a machining environment
- Describe proper housekeeping for a machining environment
- Describe the purpose of lockout/tagout procedures
- Define the terms *NFPA* and *HMIS*
- Identify and interpret NFPA and HMIS labeling systems
- Define the term *MSDS*
- Identify and interpret MSDS terms
- Interpret MSDS information
- Select the proper fire extinguisher application

Carefully read the unit, and then answer the following questions in the space provided.

Fill in the Blank

1. _____ and _____ are good practices when talking about safety.
2. _____ is the most common type of PPE used in the machining industry.
3. There are two basic types of respirators: _____ and _____.
4. A _____ or _____ may keep a machine from operating if a guard is not in its proper place.
5. When lifting an object, the knees should be _____ and the back should be _____.
6. There are two systems of labeling of hazardous materials that are widely used in the machining industry: _____ and _____.
7. The _____ is the lowest temperature where the substance can be ignited.
8. Fire extinguishers work by cooling the fuel, _____, or stopping the reaction.

Short Answer

9. What does NIOSH stand for and what is NIOSH's purpose?

_____

_____

10. List and briefly describe three types of PPE.

_____

_____

_____

11. Define the term *OSHA*.

_____

_____

12. What should anyone do with his or her jewelry when working around machinery?

_____

_____

13. List five good housekeeping rules.

_____

_____

_____

_____

_____

14. Describe the OSHA requirement "logout and tagout."

_____

_____

15. Define and briefly describe LEL and UEL.

_____

_____

16. List the three factors needed for fire to start.

_____

_____

_____

True or False

17. Different machining operations require different safety precautions.   _____

18. Many accidents are caused by not paying attention and becoming distracted.   _____

19. Safety is the responsibility of everyone.   _____

20. You should always approach someone from the front if they are operating a machine.   _____

21. OSHA requires some type of control or hearing protection if sound levels are above
115 db (decibels) for 1/4 hour or over 90 db for 8 hours.   _____

22. For those who wear regular prescription eyeglasses, safety glasses are not required.   _____

23. Compressed air can be used safely to remove metal chips, oils, solvents, or other chemicals
from machinery.   _____

24. Hazardous materials can take the form of liquids, solids, or gases.   _____

Identification Matching

25. Match the class of fire to correct combustible material.

   a. Class A   _____   **1.** Oils, gasoline, some paints, lacquers, and grease

   b. Class B   _____   **2.** Combustible metals—magnesium, titanium, and potassium

   c. Class C   _____   **3.** Paper, cloth, wood, rubber, and many plastics

   d. Class D   _____   **4.** Wiring, fuse boxes, and energized electrical equipment

Multiple Choice

26. When approaching another person who is operating a machine tool, you should always
   a. approach from behind.
   b. make a distracting movement.
   c. make a distracting sound.
   d. approach from the front.

27. The main difference between OSHA and NIOSH is
   a. one is state and the other is federal.
   b. one has limited power.
   c. NIOSH only works with small businesses.
   d. one enforces regulation and the other works on prevention of accidents.

28. The term "PPE" refers to
   a. Personal Property Equipment.
   b. Personal Production Equipment.
   c. Personal Protective Equipment.
   d. Personal Product Equipment.

29. The most common personal safety equipment used in a machine shop is
   a. steel-toed shoes.
   b. safety glasses.
   c. long sleeve shirt.
   d. shop apron.

30. The term "Housekeeping" refers to
   a. the ability to clean house.
   b. a janitor type position.
   c. cleaning up only when major spills occur.
   d. keeping the work area clean and tidy.

31. The best piece of equipment or tool(s) to use when removing chips and solvent oil from a machine is(are)
   a. compressed air with limit flow nozzle.
   b. your hands using gloves.
   c. brush and rag.
   d. scraper and pan.

32. If a small foreign particle enters the eye, what steps should be taken?

    a. Rub the eyelid to remove the particle.

    b. Gently pull the eyelid and flush with water.

    c. Remove the particle with a cotton swab.

    d. Sterilize eye area with distilled water and remove with cotton swab.

33. Before any repair or maintenance is performed on a machine, the OSHA standard of

    a. lockout/tagout should be followed.

    b. notifying the manufacturer should be followed.

    c. reporting of downtime should be followed.

    d. NFPA should be followed.

34. Companies rely on this document to identify hazardous material used in manufacturing.

    a. STEL

    b. NIOSH

    c. MSDS

    d. HMIS

SECTION 2: Measurement, Materials, and Safety

UNIT 2 — Measurement Systems and Machine Tool Math Overview

Name: _____   Date: _____

Score: _____   Text pages 63–81

Learning Objectives

After completing this unit, you should be able to:

- Understand English and metric (SI) measurement systems and perform conversions between the two
- Demonstrate understanding of fractional and decimal math and conversions between fractions and decimals
- Demonstrate ability to solve formulas and equations using basic algebra
- Identify and use properties of basic geometry
- Demonstrate understanding of angular relationships
- Perform conversions between angular measurements in decimal degrees and degrees, minutes, and seconds
- Perform addition and subtraction of angular measurements
- Demonstrate ability to locate and identify points in a Cartesian coordinate system
- Demonstrate ability to use the Pythagorean theorem
- Demonstrate the ability to solve right triangles using sine, cosine, and tangent trigonometric functions

Carefully read the unit, and then answer the following questions in the space provided.

Fill in the Blank

1. There are two types of systems of measurement: _____ and _____.
2. Uses of _____ and _____ in machining include converting units or mixing concentrates with water for cutting fluids.
3. The _____ can be labeled as _____ and is the longest side of a right triangle.
4. A complementary angle is the result of a given angle subtracted from _____ degrees, while a supplementary angle is the result of a given angle subtracted from _____ degrees.
5. The term _____ refers to a line, circle, or arc that touches a circle or an arc at only one point.

Short Answer

6. List the most common mathematical operations and the order in which they are performed.

_____

_____

7. What is the Cartesian coordinate system?

_____

_____

8. What are the three sides of a right triangle and how do the three trigonometric functions, sine, cosine, and tangent relate to the sides?

_____

_____

_____

9. Define the terms *radius* and *diameter* of a circle.

_____

_____

True or False

10. A fraction is one part of a whole.   _____
11. A fraction is a ration.   _____
12. A proportion is two ratios that are not equal to each other.   _____
13. Order of operations is the principle that certain mathematical operations must be done in a certain order to correctly perform calculations.   _____
14. The circumference of a circle is the distance across the center of the circle.   _____
15. A circle contains 360 degrees.   _____
16. The sum of the angles of every triangle is always 180 degrees.   _____
17. Parallel lines or surfaces will evidentially intersect.   _____

Computation

18. Place in order the following fractions from smallest to largest: 15/16, 3/8, 7/32, 27/64, 21/32, 1/4.

_____

_____

19. What are the decimal equivalents of the fractions from the previous question? (Write answers to the nearest 0.0001".)

_____

_____

20. What is 3/16 + 9/32?

_____

_____

21. What is 5/8 + 9/16?

_____

_____

22. What is 7/16 + .005?

_____

_____

23. What is 1/2 − 1/8?

_____

_____

24. What is 1 1/4 − 5/8?

_____

_____

25. What is 3/8 − .002?

_____

_____

26. What is 1/2 × 9/16?

_____

_____

27. What is 3/4 × 1 3/8?

_____

_____

28. What is 200 ÷ 1/4?

_____

_____

29. What is 1/2 × 1.505?

_____

_____

30. What is (.75 × 116.50 − (13 − 8) × 2) / (7 × 8 − 4)?

_____

_____

31. What is the metric equivalent of 1"?

_____

_____

32. How many inches are there in a millimeter?

_____

_____

33. What is the metric equivalent of 0.575" to the nearest .01 mm?

_____

_____

34. Convert 5.635" to mm.

_____

_____

35. Convert 29.27 mm to inches.

_____

_____

36. Solve the formula $A = \dfrac{3.82 \cdot B}{C}$ for A when B = 130 and C = 5/16.

_____

_____

37. Solve the formula $A = \dfrac{B - C}{D}$ for B when A = 1/4, C = 1.25, and D = 4.500.

_____

_____

38. Convert 60.46 degrees to degrees, minutes, and seconds. Round to the nearest second.

_____

_____

39. What is the circumference of a circle that is 5" in diameter?

_____

_____

40. Find the length of side "a" of a right triangle to the nearest 0.001" if side "b" = 2.545" and "c" = 5.850".

_____

_____

41. Find the length of side "b" of a right triangle to the nearest 0.01 mm if side "a" = 6.958" and "c" = 9.250".

_____

_____

42. What is the sine of angle "A" in the triangle if side "a" = 2.500" and side "c" = 6.500"?

_____

_____

Multiple Choice

43. The order of operation when calculating a series of numbers is as follows:
    a. Parentheses, Exponents, Addition, Division, Subtraction, Multiplication
    b. Addition, Subtraction, Exponents, Multiplication, Division, Parentheses
    c. Parentheses, Exponents, Multiplication, Division, Addition, Subtraction
    d. Subtraction, Multiplication, Addition, Division, Exponents, Parentheses

44. How many millimeters are there in 1"?
    a. 2.54
    b. .254
    c. 25.4
    d. 254.4

45. The distance from one side of a circle to the other going through the center is called the
    a. cord.
    b. diameter.
    c. radius.
    d. segment.

46. How many degrees are there in a circle?
    a. 180 degrees
    b. 90 degrees
    c. 270 degrees
    d. 360 degrees

47. A complementary angle is calculated by subtracting the known angle from
  a. 180 degrees.
  b. 90 degrees.
  c. 270 degrees.
  d. 60 degrees.

48. The Cartesian coordinate systems use two- and three-dimensional location axes, which are divided into _____ quadrants.
  a. one
  b. two
  c. three
  d. four

49. In right triangle calculations using trigonometry, the longest side of a triangle is called the
  a. adjacent side.
  b. opposite side.
  c. hypotenuse.
  d. 90-degree angle.

50. SINE is a ratio between the lengths of the
  a. opposite side and hypotenuse.
  b. hypotenuse and angle.
  c. adjunct side and opposite side.
  d. adjacent and hypotenuse.

UNIT 3	Semi-Precision Measurement

Name: _____ Date: _____

Score: _____ Text pages 82–97

Learning Objectives

After completing this unit, you should be able to:

- Define comparative measurement
- Demonstrate understanding of care of common semi-precision measuring instruments
- Read an English rule to within 1/64 of an inch
- Read an English (decimal) rule to within 1/100 of an inch
- Read a metric rule within 0.5 mm
- Identify and explain the uses of semi-precision calipers
- Identify and explain the uses of squares
- Identify and explain the uses of the combination set
- Identify and explain the uses of protractors
- Read protractors within 1 degree
- Identify and explain the uses of common semi-precision fixed gages

Carefully read the unit, and then answer the following questions in the space provided.

Fill in the Blank

1. The amount of deviation or allowable variation in part size and shape is called _____.
2. Semi-precision measurement usually refers to measurement when tolerances or levels of desired accuracy are within _____.
3. Divisions or spaces on a rule are called _____.
4. Metric rules are graduated in _____ and _____.
5. A center head is found on a _____.
6. The quickest nonprecision tool to measure a corner radius is a _____.

Short Answer

7. Briefly explain what is the most important factor when measuring with a steel rule.

   _____

   _____

8. Briefly explain what is the most important factor when measuring with a spring caliper.

   _____

   _____

9. Explain the use of a square.

_____

_____

True or False

10. Tolerances determine the type of measuring tool to be used for inspection.   _____

11. Even if one dimension is not within its specified tolerance, the part will be unacceptable and useless.   _____

12. Scales and rules are the same.   _____

13. English-reading steel rules come in fractions and decimal graduations.   _____

14. Combination bevels are a hybrid between a square and a protractor.   _____

15. Most semi-precision measuring tools do not have the ability to obtain a measurement.   _____

Identification Matching

16. Select the best semi-precision tool to determine if the given part is to print.

a. Screw pitch gage   _____   **1.** Measures parts within a 10-degree angle

b. Combination square   _____   **2.** Similar to radius gages in their function

c. Fixed gage   _____   **3.** Checks outside corner and insider corner radii

d. Angle gage   _____   **4.** Checks square, depths, and heights

e. Die maker's square   _____   **5.** Consists of two pieces: a beam and head

f. Protractor   _____   **6.** Determines the distance between threads

g. Adjustable square   _____   **7.** Measures linear distance measurement

h. Radius gage   _____   **8.** Compares angles or bevels

i. Steel rule   _____   **9.** Determines angles zero to 180 degrees

j. Bevel   _____   **10.** Specially made to check parts of certain size

Multiple Choice

17. The term semi-precision measurement usually refers to measurements when tolerances are within

    a. 1/64".

    b. 1/100".

    c. .5 mm.

    d. All of the above

18. The most common measuring tool in the machining trade is a

    a. tape measure.

    b. steel rule.

    c. yard rule.

    d. ruler.

19. Outside and inside spring calipers are considered what type of measuring tools?

    a. Direct measuring

    b. Comparison measuring

    c. Indirect measuring

    d. Transfer measuring

20. Adjustable squares are used to check for

    a. perpendicularity.

    b. squareness.

    c. flatness.

    d. All of the above

21. The best tool to check for an inside radii contour is

    a. radius gage.

    b. circle gage.

    c. fillet gage.

    d. ball gage.

SECTION 2: Measurement, Materials, and Safety

UNIT 4 | Precision Measurement

Name: _____   Date: _____

Score: _____   Text pages 98–146

Learning Objectives

After completing this unit, you should be able to:

- Explain the care of precision measuring tools
- Identify and explain the use of common precision fixed gages
- Explain the principle of the micrometer
- Identify the parts of an outside micrometer caliper
- Describe the process of outside micrometer caliper calibration
- Identify and describe uses of micrometer-type measuring tools
- Read an English micrometer
- Read a metric micrometer
- Identify and describe uses of vernier measuring tools
- Read English vernier scales
- Read metric vernier scales
- Read a vernier bevel protractor
- Identify and explain uses of precision transfer-type measuring instruments
- Identify features of dial indicators and explain their uses
- Explain the purpose of a surface plate
- Identify gage blocks and their uses, and calculate gage block builds
- Identify and explain the uses of simple and compound sine tools
- Discuss methods for measuring surface finishes
- Identify and discuss the use of a toolmaker's microscope
- Identify and discuss the use of an optical comparator

Carefully read the unit, and then answer the following questions in the space provided.

Fill in the Blank

1. A small hole gage requires a _____ to obtain a decimal measurement.
2. A _____ is used to get more accurate angle measurements than a protractor.
3. The most accurate measurement of Ra surface roughness can be obtained with a _____.
4. The best transfer measuring tool to measure a slot is _____.
5. On tools with a 25-part vernier scale, each inch is divided into _____ parts measuring _____" each.

Name: _____ Date: _____

Short Answer

6. Explain the process of adjusting a micrometer that has a worn thimble thread.

_____

_____

7. Describe the process of wringing gage blocks.

_____

_____

8. Briefly describe how to set up a 90-degree angle on a surface plate.

_____

_____

9. List three transfer-measuring tools.

_____

_____

_____

10. Explain the process of calibrating a 0"–1" micrometer and a 0"–1" depth micrometer.

_____

_____

11. How many threads per inch are on an inch-based micrometer?

_____

_____

True or False

12. Outside micrometers can be purchased that measure over 30'. _____

13. Feeler gages can also be used as shims to take up space between objects. _____

14. All vernier measuring tools have two scales on them. _____

15. Sideways movement of plunge-type indicators is acceptable. _____

16. A dial indicator looks and works much like a car's speedometer. _____

17. Adjustable parallels are used to support work in a vise. _____

Identification Matching

18. Select the best measuring tool for the given application.

a. Outside micrometer	_____	**1.** Determines flatness of a part
b. Vernier caliper	_____	**2.** A plane for measuring flatness of stock
c. Telescoping gage	_____	**3.** Measures an inside diameter of 1.75" +/−.005"
d. Straight edge	_____	**4.** Measures a round bar .500" +.001/−.002"
e. Pin gage	_____	**5.** Determines a size of hole
f. Surface plate	_____	**6.** Measures 100 pieces of 2.500" square blocks
g. CMM machine	_____	**7.** Measures a large casting with numerous critical dimensions
h. Toolmaker microscope	_____	**8.** Measures a bar 3" × 2" × .75" +/−.005"
i. Optical comparator	_____	**9.** Checks the profile of a thread
j. Plunge indicator	_____	**10.** Inspects a part .5" × .25" × .075" +/−.001"

Computations

19. If you were to make a go/no-go pin gage to check a .625" dia. $+.002"/-.003"$ hole, what would be the size of the gage? Provide a three-decimal answer.

_____

_____

20. Determine the best gage block buildup for a 36-degree angle using a 10" sine bar. Refer to pg. 108, Figure 2.4.25 in textbook as reference for block sizes in the 88-piece set.

_____

_____

21. Calculate the angle if the gage block buildup is 2.3568" on a 10" sine bar.

_____

_____

22. Record the following vernier scale readings.

   a. _____

   b. _____

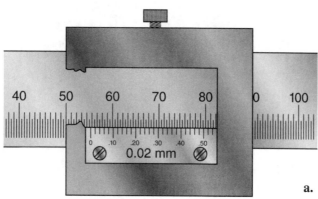

a.

©Cengage Learning 2012

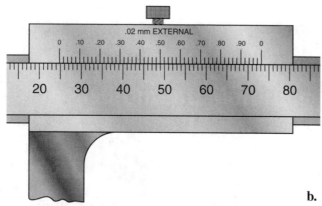

b.

©Cengage Learning 2012

Name: _____ Date: _____

23. Record the following micrometer readings.

a. _____

b. _____

c. _____

©Cengage Learning 2012

©Cengage Learning 2012

©Cengage Learning 2012

24. Record the following depth micrometer readings.

a. _____

b. _____

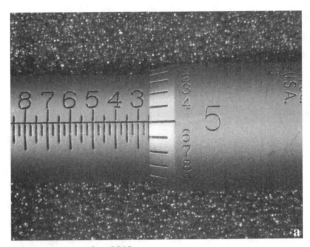

©Cengage Learning 2012

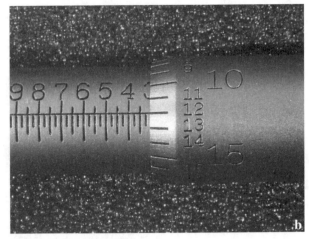

©Cengage Learning 2012

Multiple Choice

25. Precision measurement is concerned to tolerances of _____ or smaller.

    a. .010"

    b. .001"

    c. .100"

    d. All of the above

26. When determining a gage class of tolerance and accuracy, a series of numbers are given to rank the accuracy. What is the most accurate gage class listed below?

    a. Class Z

    b. Class X

    c. Class XXX

    d. Class ZZ

27. Which gage is the best for checking unthreaded external diameters (OD)?

    a. Taper plug gage

    b. Ring gage

    c. Gage blocks

    d. Threaded snap gage

28. The best precision tool to check for flatness and squareness of a part is a

    a. combination square.

    b. surface plate.

    c. beveled solid square.

    d. thick steel rule.

29. Gage blocks used in a machine shop are usually graded as

    a. Grade 0.

    b. Grade AS-1.

    c. Grade 00.

    d. Grade AS.

30. On tools with a 50-part vernier scale, each inch is broken into _____ parts measuring .050" each.

    a. 25

    b. 50

    c. 20

    d. 30

31. The best way to calibrate a standard 1"–2" micrometer is to use a(n)

    a. inside micrometer.

    b. digital caliper.

    c. adjustable parallel.

    d. gage bock.

32. The angle of the indicators' contact point is very important during setup to prevent

    a. full travel of the indicator arm.

    b. body contact of the indictor against the workpiece.

    c. indicator error in readings.

    d. All of the above

    e. None of the above

33. The center-to-center distance of the sine bar roll is representative to which side of a right triangle?

    a. Hypotenuse

    b. Opposite side

    c. Adjacent side

    d. Sine angle

SECTION 2: Measurement, Materials, and Safety

UNIT 5	Quality Assurance, Process Planning, and Quality Control

Name: _____   Date: _____

Score: _____   Text pages 147–156

Learning Objectives

After completing this unit, you should be able to:

- Define quality assurance
- Discuss the purpose of a process plan and describe its major parts
- Define and discuss the purpose of quality control
- Discuss the purpose of an inspection plan and describe its key points
- Define *SPC* and its purpose
- Identify and discuss the features of X-bar and R-charts

Carefully read the unit, and then answer the following questions in the space provided.

Fill in the Blank

1. A _____ is required to track processes and errors in the production process.
2. The first task in developing a process plan is to identify the _____
   _____.
3. The _____ determines the selection of measuring tools or equipment.
4. When a machining process is producing varying sizes of parts, the process must be _____ and adjusted or corrected.

Short Answer

5. An R-chart line is continually increasing. What is happening to this operation?

   _____

   _____

6. An R-chart line is continually decreasing. What does this tell about the operation?

   _____

   _____

7. An X-bar chart is showing a particular machined size is starting to move toward the LCL. Briefly explain what is happening and what steps should be taken.

   _____

   _____

8. What are the starting and ending steps of a process plan?

_____

_____

9. Explain the term "critical dimension" and how they affect the process planning stage.

_____

_____

10. What is the difference between QA and QC?

_____

_____

True or False

11. Quality Assurance is the plan that guides the action and performance of the machining process.   _____

12. All process plans are important and most require improvement.   _____

13. R-charts and range charts are the same.   _____

Identification Matching

14. Match the appropriate term to the defined action.

a. Sampling plan _____	**1.** Shows trends of average size of a part
b. Inspection plan _____	**2.** Explains what measuring tools to use
c. Process plan _____	**3.** Uses smallest and largest dimensions to calculate variance
d. X-bar chart _____	**4.** Defines the steps needed to make a part
e. R-chart _____	**5.** Determines the number of parts to inspect

Computations

15. What is the average of a subgroup with part sizes of .640, .635, .642, .637, .641, .643, and .639?

_____

_____

16. Using the decimal values given in the previous question, what is the range of the subgroup?

_____

_____

Multiple Choice

17. When developing the steps needed to perform a machining operation, most companies use what is called a(n)

    a. sample plan.

    b. inspection plan.

    c. quality plan.

    d. process plan.

18. A machinist may be called upon to collect dimensional data during a production. The best way to do this is to use a(n)

    a. control plan.

    b. control chart.

    c. production chart.

    d. inspection chart.

19. An R-chart, or range chart, is used to determine the

    a. variation of each sample.

    b. actual size of the sample.

    c. tool wear.

    d. machine stress during a cut.

20. During a production run, the operator notices that the R-chart is continually increasing over five subgroups but is still under the UCL. What should the operator do?

    a. Call the supervisor.

    b. Keep watching the graph and notice differences.

    c. Shut down the machine.

    d. Adjust the cut to make the increasing line decrease.

21. The UCL and LCL are determined by the part

    a. size.

    b. material.

    c. tolerance.

    d. process plan.

22. Quality control follows a protocol of steps to inspect a part. These steps are found on a

    a. FMEA.

    b. control plan.

    c. sample plan.

    d. production plan.

SECTION 2: Measurement, Materials, and Safety

UNIT 6	Metal Composition and Classification

Name: _____   Date: _____

Score: _____   Text pages 157–174

Learning Objectives

After completing this unit, you should be able to:

- Describe the difference between ferrous and nonferrous metals
- Compare and contrast low-, medium-, and high-carbon steels
- Define an alloy and an alloying element
- Describe the differences/similarities between steel and cast iron
- Demonstrate understanding of the AISI/SAE system of classification for steels
- Demonstrate understanding of UNS classification of carbon and alloy steels
- Demonstrate understanding of AA/IADS classification of aluminum alloys
- Identify UNS designations for stainless steels
- Identify UNS designations for cast iron
- Identify UNS designations for nonferrous alloys

Carefully read the unit, and then answer the following questions in the space provided.

Fill in the Blank

1. Brass is an alloy of _____ and _____.
2. Bronze is an alloy of _____ and _____.
3. Superalloys are _____ based metals.
4. Copper, magnesium, and titanium are _____ types of metal.
5. The UNS prefix for cast iron is _____.

Short Answer

6. What is tool steel? How is it the same as alloy steel? How is it different?

_____

_____

7. What are superalloys and what are their benefits?

_____

_____

8. What is the carbon content of 7240 steel?

_____

_____

9. What does a "B" in an alloy steel designate?

_____

10. The UNS prefix "S" classifies what type of metals?

_____

_____

11. What UNS prefix classifies titanium alloys?

_____

_____

12. The UNS prefix "C" classifies what type of metals?

_____

_____

13. What UNS prefix classifies superalloys?

_____

_____

True or False

14. Cast iron is very hard and brittle.   _____

15. Magnesium alloy filings and chips are highly flammable.   _____

16. All of the metal numbering systems use similar methods for their classification.   _____

17. Malleable and ductile cast irons have the ability to flex and stretch before breaking.   _____

18. A part with a high carbon content can harden during machining.   _____

Identification Matching

19. Select the best metal for the part or tool listed.

a. Aluminum	_____	**1.** Car suspension part
b. Cast iron	_____	**2.** Handrails
c. Tool steel	_____	**3.** Jet engine part
d. Cold Rolled Steel (CRS)	_____	**4.** Food conveyor part
e. Martensitic stainless steels	_____	**5.** Aircraft support brace
f. Austenitic stainless steels	_____	**6.** Axle bushing
g. Wrought iron	_____	**7.** Knee/hip joint
h. Bronze	_____	**8.** Knife blade
i. Titanium	_____	**9.** Punch and die
j. Superalloy	_____	**10.** Base for a machine

Multiple Choice

20. The main difference between ferrous and nonferrous materials is the _____ content.

    a. silicon

    b. iron

    c. carbon

    d. zinc

21. Steels that have been altered by element bombardment are called _____ steels.

    a. wrought

    b. alloy

    c. pure

    d. carbon

22. When carbon rises to above 1.7%–4.5%, the material is known as

    a. carbon steel.

    b. tool steel.

    c. cast iron.

    d. wrought iron.

23. What is the carbon content of the following material: AISI4135?

    a. 41

    b. 35

    c. 13

    d. 45

24. Which of the stainless steels below are not magnetic?

    a. Austenitic

    b. Ferrite

    c. Martensitic

    d. None of the above

UNIT 7	Heat Treatment of Metals

Name: _____ Date: _____

Score: _____ Text pages 175–189

Learning Objectives

After completing this unit, you should be able to:

- Define and discuss common heat treatment processes
- Discuss different types of heat-treating equipment
- Describe safety procedures and PPE for heat treating
- Identify and discuss Rockwell and Brinell hardness scales
- Discuss Rockwell and Brinell hardness testing methods

Carefully read the unit, and then answer the following questions in the space provided.

Fill in the Blank

1. _____ relieves stress from steel and makes it respond better to other heat-treating operations.
2. The key factor or process that transforms high content carbon steel from soft to hard is
   _____.
3. The minimum carbon content required to harden a piece of metal is _____.
4. When small parts are fed into a hopper and heated, they then pass through a bottom opening into a quenching tank; a _____ furnace is being used.
5. The best hardness tester to test material after heat treatment is _____.

Short Answer

6. Briefly describe an atmospheric furnace and its benefit.

   _____

   _____

7. Which hardness scale uses a 10-mm carbide ball to make an indentation on a test sample?

   _____

   _____

8. What is the difference between a closed loop and an open loop hardness tester?

   _____

   _____

9. List four commonly used quenching medias.

_____

_____

_____

_____

10. What hardness tester and scale uses a 1/16" ball and 100 kg load?

_____

_____

11. What process is used on aluminum to increase hardness?

_____

_____

12. List four PPEs you should have during a heat treatment process.

_____

_____

_____

_____

True or False

13. Heat treatment is only good for ferrous metals.   _____

14. There are three different scales to the Rockwell hardness tester.   _____

15. The diameter of the ball impression is used calculate the hardness of a material. _____

16. Annealing involves a quick cooling process to soften material.   _____

17. All furnaces perform the same basic task of heating metal to required temperatures. _____

Identification Matching

18. Select the appropriate heat treatment process for the given process.

    a. Case hardening   _____   **1.** Dipped or rolled in a carbon-rich powdered material

    b. Surface hardening   _____   **2.** Returns metals to their original pre-hardened condition

    c. Nitriding   _____   **3.** Removes stresses and makes structure more consistent

    d. Tempering   _____   **4.** Outer layer soaks up carbon from another source

    e. Annealing   _____   **5.** Sealed furnace containing nitrogen gas

    f. Normalizing   _____   **6.** Will decrease the steel's hardness, increase toughness

    g. Carbonizing   _____   **7.** Only an outer layer is hardened, center remains soft

Multiple Choice

19. Direct hardening of steel can only be performed if the

    a. part contains less than 30% carbon.

    b. part contains more than 30% hydrogen.

    c. part contains more than 30% silicon.

    d. part contains more than 30% carbon.

20. Induction hardening uses
    a. charcoal to surface harden the part.
    b. liquid penetrate.
    c. electrical current.
    d. chemicals.

21. The process that returns metal to it's original hardness is
    a. annealing.
    b. normalizing.
    c. tempering.
    d. carburizing.

22. If a machinist wanted to reduce stress in a casting after machining operations are complete, what heat treatment would they use?
    a. Annealing
    b. Normalizing
    c. Tempering
    d. Nitriding

23. The diamond brale penetrator and 60 kg preload are used on what hardness tester?
    a. Brinell
    b. Shore
    c. Rockwell
    d. SPIC

SECTION 2: Measurement, Materials, and Safety

UNIT 8 | Maintenance, Lubrication, and Cutting Fluid Overview

Name: _____ Date: _____

Score: _____ Text pages 190–199

Learning Objectives

After completing this unit, you should be able to:

- Describe the importance of a routine maintenance program
- Identify different methods of machine tool lubrication
- Describe routine machine tool maintenance inspection points
- Describe the purpose of cutting fluids
- Describe common types of cutting fluids
- Describe methods of application of cutting fluids

Carefully read the unit, and then answer the following questions in the space provided.

Fill in the Blank

1. A _____ is used to adjust wear in a dovetail assembly.
2. Because of their oil content, _____ lubricate better than pure synthetics, but not as well as straight or soluble oils.
3. Machine tools transmit motion to slides by threaded shafts called _____.
4. Before any work is done on internal moving parts of a machine, _____ should be completed on the machine.
5. A filtered, recirculation system used to apply cutting fluids is called a _____ system.

Short Answer

6. List three reasons why lubricants are used on machines.

   _____

   _____

   _____

7. List five nonliquid cutting compounds.

   _____

   _____

   _____

   _____

   _____

8. What are two major concerns when using synthetic oils?

_____

_____

9. What is MQL and how does it benefit tool lubrication?

_____

_____

10. Explain the term "backlash."

_____

_____

11. Define the term "MSDS."

_____

_____

True or False

12. Preventive maintenance tasks usually need to be done at specific intervals.   _____

13. Any oil or grease will work to lubricate a machine tool.   _____

14. Chemical-based fluids are combined with water and should never be used for magnesium.   _____

15. Chemical-based cutting fluids provide the best cooling level for machining.   _____

16. A light beam can be used to determine concentration levels of mixed fluids.   _____

Multiple Choice

17. Before using any liquid solvent, grease, or oil you should
   a. read the machine manual to check recommendations.
   b. read the PM shop report.
   c. check machine levels.
   d. read the MSDS and PPE requirements.

18. Lubricants are used for
   a. cooling parts.
   b. minimizing friction.
   c. preventing seizures.
   d. All of the above.

19. Water-based cutting fluids are called
   a. soluble oils.
   b. straight oils.
   c. chemical-based.
   d. water fluids.

20. Cutting fluids that contain no oil are called

    a. soluble oils.

    b. straight oils.

    c. synthetics.

    d. chemical-based.

21. When using soluble oils and synthetic cutting fluids, it is important to check the concentration levels of the mixture. A handheld device used for this is called a

    a. fraction meter.

    b. ratio meter.

    c. refractometer.

    d. soluble meter.

UNIT 1	Understanding Drawings

Name: _____ Date: _____

Score: _____ Text pages 200–231

Learning Objectives

After completing this unit, you should be able to:

- Identify and interpret title block information
- Identify line types and their uses
- Describe the principle of orthographic projection
- Identify the three basic views frequently used in engineering drawings
- Identify isometric views
- Identify and describe the use of basic symbols and notation used on engineering drawings
- Define *tolerance*
- Demonstrate understanding of unilateral, bilateral, and limit tolerances
- Demonstrate understanding of allowances and classes of fit for cylindrical components
- Identify basic geometric dimensioning and tolerancing (GD&T) symbols
- Interpret basic GD&T feature control frames

Carefully read the unit, and then answer the following questions in the space provided.

Fill in the Blank

1. _____ _____ can be used to show alternate positions or outlines of adjacent parts.
2. A _____ _____ line can be used to make an imaginary cut through an object to create a _____ view.
3. A machine operator uses the _____ of a print to understand past changes or revisions.
4. _____ _____ is the name of the three-dimensional view that is sometimes shown on prints to help visualize an object.
5. GD&T uses a _____ _____ _____ to show how to control the size and/or shape of a particular feature of a part.

Short Answers

6. If the scale on an engineering drawing is 1:2, is the drawing larger or smaller than the actual object?

_____

_____

7. Briefly define *orthographic projection*.

_____

_____

8. Sketch the symbols used on prints to identify the following:
   a. Countersink _____
   b. Diameter _____
   c. Counterbore _____
   d. Depth _____
   e. Position _____

9. What are the symbols used for datum planes?

   _____

   _____

10. To control a shaft from wobbling, what GD&T symbol would be used on the print?

   _____

   _____

True or False

11. Radii on the inside and outside corners of a workpiece are sometimes called fillets and rounds. _____

12. General notes can be used to establish tolerances. _____

13. A basic size includes tolerances of the part. _____

14. A bonus tolerance can be given for dimensions that are in the LMC state. _____

15. If a machine operator has an issue with a dimension on the print, he or she can measure the print dimensions and get the correct answer. _____

16. Two surfaces can be parallel but not flat. _____

Identification Matching

17. Match the type of line to its description.
   a. Object _____
   b. Hidden _____
   c. Center _____
   d. Dimension _____
   e. Leader _____
   f. Extension _____
   g. Phantom _____
   h. Cutting plane _____

   1. A thin, angled line with an arrowhead on one end that points to a specific feature or detail
   2. A thin line that has arrowheads at the ends
   3. A line that shows alternate positions of a part or outlines of adjacent parts
   4. A thick and continuous line
   5. A thick line that is drawn as one long and two short dashes alternately spaced
   6. A thin line broken into a series of short dashes
   7. A line that extends from the edges of an object or feature
   8. A thin line broken into alternating long and short dashes

Computations

18. Determine the total tolerance, upper limit, and lower limit for the following dimensions:

a. $1/2 \pm \dfrac{1}{64}$

b. $3/4 + \dfrac{1}{64} / -0$

c. $5/8 + 0 / \dfrac{1}{-64}$

d. $1.000 \pm .003$

e. $1.0625 + 0 / - .0005$

f. $.625 + .001/-0$

19. What is the MMC of a shaft with a dimension of $1.250 + 0 / -.001$?

_____

20. What is the LMC of a shaft with a dimension of $1.750 + 0 / -.002$?

_____

21. What is the LMC of a hole with a dimension of $.375 \pm .001$?

_____

22. What is the MMC of a hole with a dimension of $.875 \pm .002$?

_____

23. Calculate the bilateral tolerance when given the unilateral tolerance of $2.505 +.010 / -.0000$.

_____

24. If a shaft is dimensioned at $\dfrac{1.0021}{1.0018}$ and the hole of the mating part is dimensioned at $\dfrac{1.0025}{1.0031}$, is the allowance positive or negative? What is the amount of the allowance?

Multiple Choice

25. If a machinist needed to determine the revision of the print, where should he or she look on the blueprint?

a. Body of the drawing

b. BOM list

c. Title block

d. Above the title block

26. The method of representing a three-dimensional object in a two-dimensional format is called

a. view arrangement.

b. orthographic projection.

c. single-view projection.

d. three-view projection.

27. When creating a drawing that has alternating or rotating parts, what line would be best to illustrate the motion?

a. Hidden

b. Phantom

c. Center

d. Solid

28. In order to show the internal features of a part, a _____ view is required.

    a. front

    b. side

    c. section

    d. rotated

29. Convert the following unilateral tolerance to a bilateral tolerance: 2.500" +.006"/−.000".

    a. 2.505" + .001/−.005"

    b. 2.504" + .002"/−.004"

    c. 2.503" + .003/−.003"

    d. All the above

    e. None of the above

30. The relationship between the sizes of two mating parts is called the

    a. class of fit.

    b. allowance.

    c. tolerance.

    d. interference.

31. The main difference between circularity and cylindricity is the control of the dimension within

    a. a section.

    b. the length of part.

    c. the length of the surface.

    d. the diameter.

32. The term "*MMC*" refers to

    a. minimum material control.

    b. minimum machining control.

    c. maximum material control.

    d. maximum machining condition.

SECTION 3: Job Planning, Benchwork, and Layout

UNIT 2   Layout

Name: _____   Date: _____

Score: _____   Text pages 232–249

Learning Objectives

After completing this unit, you should be able to:

- Define *layout* and explain its purpose
- Identify and use common semi-precision layout tools
- Identify and use common precision layout tools
- Perform typical mathematical calculations required to perform layout
- Perform basic layout procedures

Carefully read the unit, and then answer the following questions in the space provided.

Fill in the Blank

1. To transfer a linear dimension from one point to another, a _____ can be used.
2. A common tool used during a layout project is a _____.
3. The most precise way to scribe angles is to use a _____ and _____.
4. If a machinist were to lay out a line parallel to an edge of a workpiece, the machinist would use a _____ to perform the task.
5. A reference plane for performing a precision layout is a _____.

Short Answer

6. What is the purpose of layout?

_____

_____

7. What semi-precision tool can be used to find the center of a cylindrical piece of material?

_____

_____

8. What layout tool would be a good choice for laying out a 44" diameter?

_____

_____

9. What advantages does a height gage (vernier, dial, or digital) have over a surface gage?

_____

_____

10. What purpose do parallels and setup blocks serve?

_____

_____

11. What type of tool can be used with gage blocks and a surface plate to position work for very accurate angular layout?

_____

_____

12. List two uses of a V-block.

_____

_____

13. What is the difference between a center punch and prick punch?

True or False

14. Protractors can be used to lay out angles with an accuracy of 5 minutes (1/12 of a degree). _____

15. Semi-precision layout can be used when print tolerances are 1/32" or greater. _____

16. Height gages do not have the ability to read measurements. _____

17. Parallels are used to measure workpieces from a reference plane. _____

18. A machinist should always read the MSDS for layout ink. _____

Computations

19. What is the largest square that could be produced from 1.5"-diameter round material?

_____

20. A 1" square piece of material is needed, but only round material is available. What size piece of round material would be required to produce this 1" square?

_____

21. What is the angle of a workpiece if a 3.2003" gage block stack is used with 10" sine bar? Answer in degrees and minutes.

_____

Multiple Choice

22. Layout is used to
   a. determine the number of pieces you can get from a blank.
   b. help determine where to cut and not cut.
   c. help determine the orientation of the piece.
   d. All of the above
   e. None of the above

23. During layout, a machinist is required to create a few circles, radii, and arcs on a flat workpiece. What is the best tool for the task?
   a. Center head
   b. Combination set
   c. Trammel
   d. Dividers

24. The main difference between a prick punch and a center punch is the
    a. length.
    b. degree of the point.
    c. material.
    d. diameter.

25. The best tool to use to hold a round workpiece during layout is a(n)
    a. angle plate.
    b. vise.
    c. V-block.
    d. hand clamp.

26. Precision layout is considered when tolerances are closer than
    a. 1/64".
    b. .030".
    c. .5 mm.
    d. All of the above
    e. None of the above

UNIT 3	Hand Tools

Name: _____ Date: _____

Score: _____ Text pages 250–265

Learning Objectives

After completing this unit, you should be able to:

- Identify common hand tools
- Describe the uses for common hand tools
- Describe hand tool safety precautions

Carefully read the unit, and then answer the following questions in the space provided.

Fill in the Blank

1. When there is not enough space beyond a screw head to use a standard screwdriver, a(n) _____ can be used instead.
2. A hacksaw cuts on the _____ stroke.
3. There are _____ common cross sectional file styles a machinist can choose from.
4. If a machinist was to form a piece of light sheet metal, they could use a _____ _____ hammer to perform the task.
5. The fastest tool used to remove a nut would be a _____ wrench.
6. The automobile industry mainly uses _____ screwdrivers with a six-pline tip.
7. The narrow slot produced when sawing with a hacksaw blade is called the _____.

Short Answer

8. What is an advantage of using tongue-and-groove pliers?

_____

_____

9. What is the purpose of a dead blow hammer?

_____

_____

10. In what situation would an open-end wrench be chosen over a box-end wrench?

_____

_____

11. What is the biggest disadvantage of using an open-end wrench?

_____

_____

12. Why should there be at least three teeth in contact with the material when hacksawing?

_____

_____

13. List the four types of file tooth classifications.

_____

_____

_____

_____

14. List the six levels of file coarseness from roughest to smoothest.

_____   _____

_____   _____

_____   _____

15. What is a safe edge and what is it used for?

_____

_____

True or False

16. The larger the screwdriver, the more force you can apply to the screw.   _____

17. Box-end wrenches produce the same force to a bolt as open-end wrenches.   _____

18. A vise with hard jaws would be used for light clamping.   _____

19. A hacksaw has four main parts.   _____

20. Pinning and loading are caused by too much pressure during filing.   _____

21. Slip joint and tongue-and-groove pliers are one in the same.   _____

Identification Matching

22. Select the best hand tool from the list for the given application.

a. Form a sheet metal rivet _____	**1.** Dead blow hammer
b. Remove a screw from a car _____	**2.** Ball peen hammer
c. Clamp a large, round workpiece _____	**3.** Slip joint pliers
d. Remove a clip lodged between two surfaces _____	**4.** Diagonal pliers
e. Seat a workpiece in a vise on parallels _____	**5.** Needle nose pliers
f. Cut small wire extending from a surface _____	**6.** Torx screwdriver

Multiple Choice

23. If a machinist were to pin rivet heads that join two pieces of metal together, which would be the best hammer for the task?
    a. Brass head
    b. Ball peen
    c. Soft face
    d. Dead blow

24. When seating a workpiece in a vise on parallels, which is the best hammer for the task?
    a. Brass head
    b. Ball peen
    c. Soft face
    d. Dead blow

25. Your task is to remove nuts from bolts that have been in place for many years; the heads are corroded and dirty. Which is the best wrench to remove the nuts from the bolts?
    a. Socket wrench
    b. Box-end wrench
    c. Open-end wrench
    d. Adjustable wrench

26. When using an adjustable wrench the wrench, is rotated in the direction of the
    a. adjustable jaw.
    b. solid jaw.
    c. wrench handle.
    d. center line of the bolt head.

27. A hacksaw always has the blade teeth pointing
    a. forward, away from operator.
    b. toward the material.
    c. away from the handle.
    d. All of the above
    e. None of the above

28. The narrow cut in the workpiece produced by the hacksaw blade is called the
    a. kerf.
    b. set.
    c. rake.
    d. blade thickness.

29. Usually, the length of the file determines the
    a. coarseness.
    b. amount of material to be removed.
    c. single or double cut.
    d. All of the above
    e. None of the above

30. The proper amount of pressure is required during filing to prevent
    a. loading.
    b. pinning.
    c. workpiece scratches.
    d. All of the above
    e. None of the above

UNIT 4	Saws and Cutoff Machines

Name: _____ Date: _____

Score: _____ Text pages 266–282

Learning Objectives

After completing this unit, you should be able to:

- Identify the various sawing machines used in the machine shop
- Operate band saws safely
- List the different band saw blade materials
- Understand blade pitch
- Identify the three different tooth patterns and their uses
- Identify the three different blade sets and their uses
- Describe how to select proper band saw blade width
- Understand and be able to identify saw tooth geometry
- Explain the term *kerf*
- Calculate band saw blade length
- Describe the band welding procedure
- Describe blade mounting procedure for the vertical band saw

Carefully read the unit, and then answer the following questions in the space provided.

Fill in the Blank

1. Soft materials such as aluminum are commonly cut with the _____ set saw blade.
2. The main advantage of the abrasive or chop saw is that it can cut _____ _____.
3. A cold saw can use a _____ or a _____ _____ circular blade.
4. The most important considerations a machinist must consider when preparing to perform a cutting operation on a band saw is _____ _____.
5. A large, round steel workpiece or large piece of steel flat stock would commonly be cut with a _____ tooth set.
6. The harder the material, the _____ the speed of the blade.

Short Answer

7. What is the major disadvantage of the power hacksaw?

_____

_____

8. What can be done to make turning corners easier when contour sawing?

_____

_____

9. What is a variable-pitch saw blade, and what are its benefits?

_____

_____

10. What is the difference between blade gage and blade set?

_____

_____

11. What is the main difference between the standard tooth blade and the skip tooth blade?

_____

_____

12. List the three common tooth sets.

_____

_____

_____

True or False

13. A band saw's blade performance depends on its pitch and tooth pattern.   _____

14. The speed of a band saw blade is the distance that 1 foot of blade would travel in 1 minute.   _____

15. Abrasive saws have very little blade "walking," or flexing.   _____

16. Positive-rake teeth are generally thinner and weaker than negative-rake teeth.   _____

Computations

17. What minimum number of teeth per inch should be used to saw 3/16"-thick material?

_____

18. What length of blade stock would be needed for a saw with 24"-diameter wheels that are 36" on center?

_____

19. What minimum width of saw blade would be needed to saw a contour with a 1/2" radius?

_____

20. What is the approximate thickness of material if you had a 12 tooth per inch blade?

_____

Multiple Choice

21. The main difference between a vertical and horizontal band saw is the
    a. type of blade they use.
    b. type of material they can cut.
    c. orientation of the band in the saw.
    d. degree of accuracy of the cut.

22. The main difference between a flex-back and hard-back blade is the
    a. teeth alignment.
    b. set and rake of the teeth.
    c. width of the blade.
    d. hardness of the blade.

23. The alternate tooth set pattern has teeth that
    a. switch from side to side every other tooth.
    b. are intermittent, with a neutral space between.
    c. are grouped in a wavy pattern.
    d. None of the above

24. The metal chip is formed into curls when sawing at what part on the blade?
    a. Tooth tip
    b. Set gap
    c. Gullet
    d. Crest

25. Speeds for band sawing machines are given in
    a. SFPM.
    b. IPM.
    c. SIPM.
    d. RPM.

26. As a rule of thumb, the harder the material, the _____ speed.
    a. faster
    b. slower
    c. more
    d. lower

SECTION 3: Job Planning, Benchwork, and Layout

UNIT 5 | Offhand Grinding

Name: _____ Date: _____

Score: _____ Text pages 283–292

Learning Objectives

After completing this unit, you should be able to:

- Identify uses of offhand grinding
- Select the correct grinding wheel for the operation to be performed
- Identify different types of offhand grinding machines
- Install and dress a grinding wheel
- Set up a pedestal grinder for safe operation
- Safely perform offhand grinding

Carefully read the unit, and then answer the following questions in the space provided.

Fill in the Blank

1. The wheel blotter has printed on it the maximum _____ of the wheel.
2. To be sure that a grinding wheel did not sustain _____ during removal, storage, or transport, a machinist must always perform a _____ prior to mounting a wheel.
3. Too much pressure on the part moving into the grinding wheel will cause _____ of the wheel.
4. A commonly used grinding wheel is the _____ wheel.
5. _____ wheels are used for grinding the extremely hard carbides used in cutting tools as well as non-ferrous metals.

Short Answer

6. List the three types of abrasives and the types of materials that should be ground by each abrasive type.

   _____   _____

   _____   _____

7. Briefly define a loaded grinding wheel.

   _____

   _____

8. Briefly define a glazed grinding wheel.

   _____

   _____

9. Briefly list five offhand grinding safety precautions.

   _____   _____

   _____   _____

   _____   _____

10. List three considerations when determining a grinding wheel size.

_____

_____

_____

11. Why do you stand to one side of the pedestal grinder during start-up?

_____

_____

12. During a pedestal grinding operation, a small fire started because of sparks. What was wrong or could have been done to prevent this?

_____

_____

True or False

13. A 60-grit grinding wheel is coarser than a 120-grit grinding wheel.   _____

14. Grinding wheels are made to be tough and long-lasting.   _____

15. The minimum gap for a tool rest should be 1/16.   _____

16. Star and the wheel dressers are one and the same.   _____

17. Pedestal grinders can be used for grinding ferrous and non-ferrous materials.   _____

18. Grinding wheels remove less material than abrasive belts.   _____

Multiple Choice

19. Before mounting a new grinding wheel on a grinder, what should be done first?
   a. Check wheel color
   b. Check grinder safety guards
   c. Apply grease to the bearings
   d. Perform a ring test on the wheel to be mounted

20. Tool rest and spark breaker should be adjusted to a clearance of
   a. .0625" from the wheel.
   b. .100" from the wheel.
   c. .0325" from the wheel.
   d. as close as possible to the wheel.

21. The process of removing loading and glazing of a grinding wheel is called
   a. pinning.
   b. dressing.
   c. filing.
   d. truing.

22. The most common type of grinding wheel abrasive material is
   a. white silicon carbide.
   b. diamond impregnated.
   c. aluminum oxide.
   d. green silicon.

SECTION 3: Job Planning, Benchwork, and Layout

UNIT 6	Drilling, Threading, Tapping, and Reaming

Name: _____   Date: _____

Score: _____   Text pages 293–309

Learning Objectives

After completing this unit, you should be able to:

- Demonstrate understanding of benchwork drilling operations
- Demonstrate understanding of countersinking, spotfacing, and counterboring
- Identify various reamer types and explain their use
- Demonstrate understanding of standardized thread systems and their designations
- Identify various tap types and explain their use
- Demonstrate understanding of tap drill selection
- Identify various thread-cutting die types and explain their use
- Demonstrate understanding of tap removal techniques

Carefully read the unit, and then answer the following questions in the space provided.

Fill in the Blank

1. A thread-cutting tool called a _____ is used to create external threads by hand.
2. _____ evacuate chips from the hole while the drill is rotating.
3. Before drilling, a machinist should _____ the center of the hole.
4. _____ is the drilling operation that creates a smooth bearing surface.
5. Before tapping a hole, a _____ should be used on the hole to help guide/start the tap.
6. When reaming deep holes, a _____ reamer is recommended.
7. A _____ is used to hold threading dies when cutting threads.

Short Answer

8. What are the two main standardized systems of threads used in benchwork?

_____

_____

9. What is the pitch of a 1/2-20 thread?

_____

_____

10. What are the three taps in a standard tap set (based on chamfer type)?

_____

_____

_____

11. In which order should the taps in question 10 be used when tapping a blind hole?

_____

_____

_____

12. What is the concern when a drill bit goes through the metal?

_____

_____

13. Write a detailed machine process plan of hand reaming a .250" diameter hole.

_____

_____

_____

_____

_____

14. List the four methods of identifying twist drill diameters.

_____

_____

15. Define the term UNF 1/4"-28 2A.

_____

_____

True or False

16. Some countersinks have pilots. _____

17. The bigger the hole to be hand reamed, the more material is left in the hole. _____

18. Reamer RPM should be the same as in spot drilling. _____

19. The lead of a thread is the distance from a point on one thread to the same point on the adjacent thread. _____

20. Twist drills produce a larger hole than the diameter of the drill itself. _____

21. The bigger the hole diameter, the less the bolt percent of thread strength. _____

Identification Matching

22. Identify the following thread parts by putting the corresponding letter in the appropriate circle on the diagram below.

    a. Major diameter

    b. Minor diameter

    c. Root

    d. Crest

    e. Pitch diameter

    f. Included angle

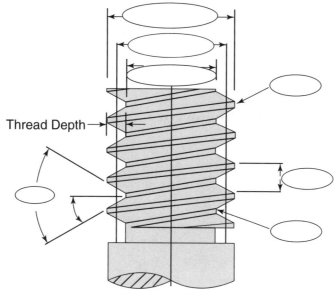

Thread Depth →

©Cengage Learning 2012

Computations

23. Using the tap drill chart within this unit, determine the tap drill sizes for the following threads:

    a. 3/8-16 UNC   _____

    b. 1/4-28 UNF   _____

    c. 10-24 UNC   _____

    d. 6-40 UNF   _____

    e. 1/2-13 UNC   _____

    f. 3/8-24 UNF   _____

Multiple Choice

24. A machining process that creates a flat spot on a rough surface is

    a. countersinking.

    b. spotfacing.

    c. counterboring.

    d. chamfering.

25. A machining process that creates an included "V" angle or taper in a hole is

   a. countersinking.
   b. spotfacing.
   c. counterboring.
   d. chamfering.

26. Reamers are used to

   a. create a larger hole.
   b. rough out a hole to size.
   c. create a small hole.
   d. All of the above
   e. None of the above

27. When reaming, the proper amount of material for a hole up to 1/4" is

   a. >.010".
   b. <.010".
   c. >.015".
   d. <.015".

28. What is the series of the following thread: 3/4-10 UNC -3A?

   a. 3A
   b. 3/4"
   c. UNC
   d. 10

29. Most tap drill charts suggest a hole size that will produce about _____ thread depth.

   a. 60%
   b. 50%
   c. 80%
   d. 75%

30. Which tap is considered a general, all-purpose tap?

   a. Bottoming tap
   b. Plug tap
   c. Taper tap
   d. Spiral tap

SECTION 4: Drill Press

UNIT 1 | Introduction to the Drill Press

Name: _____ Date: _____

Score: _____ Text pages 310–318

Learning Objectives

After completing this unit, you should be able to:

• Identify types of drill presses
• Identify the major components of the drill press and their functions

Carefully read the unit, and then answer the following questions in the space provided.

Fill in the Blank

1. Upright drill presses are available in _____ and _____ models.
2. The base and column of an upright drill press are mounted _____ degrees from each other.
3. Upright drill presses are available with either-driven _____ or _____ driven-heads.
4. There are _____ types of spindle-driven upright drill presses.
5. The _____ is the largest of the upright drill presses.
6. On an upright drill press, the _____ or _____ holds the cutting tool.

Short Answer

7. List two general types of upright drill presses.

   _____

   _____

8. List two disadvantages of a belt-driven upright drill press.

   _____

   _____

9. Describe the advantage of using a multi-spindle upright drill press.

   _____

   _____

10. Briefly describe the term "sensitive" drill press.

   _____

   _____

11. Briefly describe a micro drill press and its purpose.

_____

_____

12. What would be the best drill press to machine holes that are 1" or larger in diameter?

_____

_____

13. What would be the best drill press to machine holes as small as .005" diameter?

_____

_____

14. List five common machining operations that can be performed on an upright drill press.

_____

_____

_____

_____

_____

_____

True or False

15. The size of an upright drill press is determined by the distance from the center of the spindle to the outside diameter of the column.   _____

16. The best upright drill press to perform delicate work on large parts is a sensitive drill press.   _____

17. The spindle should always be turning when selecting speeds on a gear-head upright drill press.   _____

18. The multi-spindle drill press is also called the gang drill press.   _____

19. On a radial-arm drill press, the head moves over the secured workpiece.   _____

20. The table of an upright drill press is always mounted to the base.   _____

21. The tool used to secure a cutting tool in a chuck is called a key.   _____

22. Spindle speed is measured in FPM, or feet per minute.   _____

Identification Matching

23. Match the following upright drill press parts to the function they perform.

a. Column	_____	1. Contains the drive motor
b. Spindle	_____	2. Automatically advances the cutting tool
c. Speed selector	_____	3. Feeds the cutting tool into the workpiece
d. Quill	_____	4. Supports the worktable and head
e. Depth stop	_____	5. Holds the cutting tool
f. Table clamp	_____	6. Adjusts the spindle speed
g. Head	_____	7. Secures and locks the table
h. Power feed	_____	8. Limits quill travel

24. Identify the following upright drill parts by putting the corresponding letter in the appropriate circle on the diagram below.

- **a.** Worktable
- **b.** Head
- **c.** Base
- **d.** Column
- **e.** Speed selector
- **f.** Quill feed handle
- **g.** Spindle
- **h.** Depth stop
- **i.** Forward/Off/Reverse switch
- **j.** Elevating crank
- **k.** Table clamp
- **l.** Quill

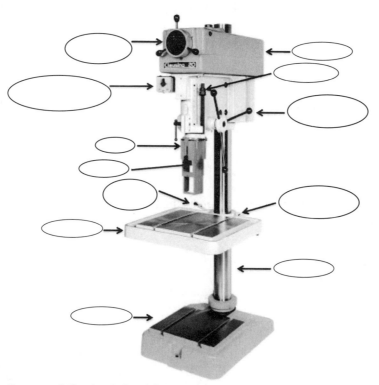

Courtesy of Clausing Industrial, Inc.

Multiple Choice

25. The spindle speed of a drill press is expressed in what unit?

    a. FPM

    b. RPM

    c. IPM

    d. SFPM

26. A drill press that does not have an automatic feed mechanism is sometimes referred to as a(n)

    a. upright drill press.

    b. bench drill press.

    c. radial-arm drill press.

    d. sensitive drill press.

27. The size of an upright drill press is determined by the

    a. largest size of drill the spindle can hold.

    b. size of the motor.

    c. largest-diameter workpiece that can be drilled on center.

    d. length from the center of the table to the center of column.

28. What drill press would the best choice to perform drilling on a large, irregularly shaped casting?

    a. Gang drill press

    b. Radial-arm drill press

    c. Magnetic drill dress

    d. Sensitive drill press

29. The part of an upright drill press that contains the spindle and is moved by the feed handle is called the

    a. quill.

    b. quilt.

    c. spindle housing.

    d. spindle sleeve.

30. Micro precision drill presses are only capable of producing holes

    a. >.250" in diameter.

    b. <.250" in diameter.

    c. that need accurate location.

    d. in soft material.

UNIT 2	Tools, Tool Holding, and Workholding for the Drill Press

Name: _____ Date: _____

Score: _____ Text pages 319–334

Learning Objectives

After completing this unit, you should be able to:

- Identify the major parts of the twist drill
- Explain the function of each part of the twist drill
- Explain the various tool holding and workholding devices used on the drill press
- Identify which type of tool holding and workholding device should be used in various situations

Carefully read the unit, and then answer the following questions in the space provided.

Fill in the Blank

1. The reaming operation is intended to produce very _____ holes with exceptional _____.

2. The _____ of the straight reamer produces the cutting action.

3. _____ is a cutting-tool material capable of superior life due to its extreme hardness, greater toughness, wear resistance, and heat resistance.

4. _____ holes are most often used in the casting processing industry.

5. There are _____ points of contact on the workpiece when using a V-block.

6. A _____ is a micro drill chuck that will enable tiny drill bits to run truer than if they were placed in larger chucks.

7. _____ can be used to secure work to a table at a 90-degree position.

Short Answer

8. What is a common material drill bits are made from?

_____

_____

9. Sketch an illustration of a counterbore. Show both the bolt and the workpiece.

10. Sketch an illustration of a spotface. Show both the bolt and the workpiece.

11. What are the two types of drill shaft shanks?
_____
_____

12. What type of drill bit shank is likely to be found on a 1 1/2"-diameter twist drill?
_____
_____

13. Name the three primary sections of the drill bit.
_____
_____
_____

14. Briefly list each of the steps required to secure the setup of a workpiece in a drill press vise using parallels.
_____
_____
_____
_____
_____

15. What are the two main reasons to use a countersink?
_____
_____

16. Briefly explain the use of pilot.
_____
_____

17. Briefly explain the use of a drill drift.
_____
_____

True or False

18. Steel bars called parallels can be used to support the workpiece and provide clearance between the table. _____

19. The dead center of the drill is what forces the material to the flute, allowing it to be cut by the lips. _____

20. Drill with diameters larger than 1/2" must have a straight shank. _____

21. Countersinks are available with many different included angles. _____

22. A Morse taper sleeve reduces the size of a tool's shank taper. _____

Identification Matching

23. Match the following twist drill body parts to the their appropriate function.

a. Lips   _____   1. Tapered portion that makes up the centermost part of the drill

b. Chisel edge   _____   2. Provides area for mounting the drill bit

c. Flutes   _____   3. Pathway for chips and coolant

d. Margin   _____   4. Angle of the spiral relative to the center axis

e. Body clearance   _____   5. Only part of the drill that does cutting

f. Helix angle   _____   6. Prevents rubbing of the drill in the hole

g. Web   _____   7. First point of contact into the workpiece

h. Shank   _____   8. Gives a drill bit its diameter

24. Match the workpiece to the appropriate workholding device or apparatus.

a. Drill press vise   _____   1. 1/2" × 4" × 6" flat plate

b. V-block   _____   2. 1/2" diameter × 6" long CRS bar

c. Angle plate   _____   3. 1" × 3" × 4" tool steel, irregular contour

d. Hold-down clamps   _____   4. 4" × 4" × 6" cast-iron casting

Multiple Choice

25. The cutting action of a twist drill can be located on what part of the drill?

a. Dead center

b. Heel

c. Flutes

d. Lips

26. After inspection of a hole diameter, it was found that the diameter of the produced hole was larger than the drill size by .010". What would be the cause?

a. Too little feed

b. RPM to high

c. Drill not sharp

d. Too hard of material

27. The included angle of a typical twist drill is

a. 130 degrees.

b. 60 degrees.

c. 82 degrees.

d. 118 degrees.

28. The part of the twist drill that gives the drill its diameter is called the

a. body.

b. flute.

c. margin.

d. web angle.

29. Using the Morse taper numbers given, which of the following drills is larger?

    a. 4

    b. 5

    c. 6

    d. 7

30. Identify the part of the reamer that actually does the cutting of the material.

    a. Lips

    b. Flute

    c. Chamfer

    d. Body

31. The best type of drill chuck to hold very small drills is the

    a. sensitive drill chuck.

    b. pin vise chuck.

    c. Jacob #1.

    d. keyless chuck.

32. A V-block offers more holding power than a vise because of the

    a. clamp.

    b. V-shaped design.

    c. three contact areas.

    d. greater force that can be applied.

UNIT 3 | Drill Press Operations

Name: _____ Date: _____

Score: _____ Text pages 335–351

Learning Objectives

After completing this unit, you should be able to:

- Describe drill press safety procedures
- Define cutting speed and perform speed and feed calculations for holemaking operations
- Explain procedures for drilling operations
- Explain procedures for reaming operations
- Explain procedures for countersinking operations and calculate countersink feed depth
- Explain procedures for counterboring/spotfacing operations
- Explain procedures for tapping operations and estimate number of tap turns to achieve a given thread depth

Carefully read the unit, and then answer the following questions in the space provided.

Fill in the Blank

1. When drilling holes over 1/2" in diameter, a _____ _____ should first be drilled with a smaller drill.

2. A hole that is only drilled partially through a workpiece is called a _____ hole.

3. A _____ _____ can be used to power tap holes on the drill press and will automatically reverse the tap and retract it from the hole when the quill is raised.

4. A _____ _____ is used to control the depth of a drilling operation on a drill press.

5. The spot or center drill depth should be _____ the length of the angled point of the twist drill being used.

6. Feed rates that are too _____ can cause rapid tool wear and even tool breakage.

7. The harder the material, the _____ the spindle speed.

8. The larger the diameter of the cutting tool, the _____ the spindle speed.

Short Answer

9. Briefly explain feed rate and how different metals affect the feed rate selection.

_____

_____

10. List the four basic factors that will influence the RPM required to perform any hole-making operation.

_____   _____

_____   _____

11. If a part has a machinability rating of 55%, will special machining be required? Explain.

_____

_____

12. Describe the process of locating a hole center using a wiggler.

_____

_____

13. How does Brinell hardness relate to SFPM?

_____

_____

True or False

14. When machine reaming a hole, if too little material is left for finishing, the reamer
may not cut properly.   _____

15. A good practice is to always bottom the reamer in a hole to ensure a complete diameter.   _____

16. The angle of a drill point is included when specifying the depth of a hole.   _____

17. Drill press feed rate is measured in FPR.   _____

18. A machinist should always set-up the drill press so the long side of the workpiece
is to the right of the hole being drilled.   _____

19. Tapping can be performed on the drill press with a tapping head.   _____

Computations

20. Calculate spindle RPM for the operation of drilling a 13/16" hole in CRS that has a 90-SFPM cutting speed.

_____

21. Calculate feed depth for a 90-degree countersink used to machine a 3/4"-diameter countersink on an existing
3/8"-diameter hole.

_____

22. If a tapped 1/4-20 hole requires a measured thread depth of .4375", how many turns of the tap are required?

_____

23. If the spindle speed is 1,200 RPM and the cutting speed of the metal is 120 SFPM, what is the estimated
size of the drill?

_____

Multiple Choice

24. If a piece of material has a machinability rating of 70%, this would mean the material
  a. is going to be difficult to machine.
  b. is going to be easy to machine.
  c. can be removed at 70% of the standard rate.
  d. can be removed 70% faster than at the standard rate.

25. Cutting speed is the distance that a point on the circumference of a rotating cutting tool travels in 1 minute. It is stated in what unit?

    a. SFPM

    b. SFM

    c. FPM

    d. All of the above

    e. None of the above

26. Drill press feed rates are expressed in what unit?

    a. IPR

    b. SFPM

    c. RPM

    d. MPR

27. What is the RPM used for a piece of material with a CS of 90 and a drill with a diameter of 1/2"?

    a. 600 RPM

    b. 687 RPM

    c. 500 RPM

    d. 725 RPM

28. The best tool to use to find the center of pre-existing hole is a

    a. wiggler.

    b. center finder.

    c. pointed edge finder.

    d. All of the above

    e. None of the above

29. A good rule of thumb for spot drill depth is that it should be

    a. 3/4 the length of the drill chamfer.

    b. 1/2 the length of the drill point.

    c. 1/2 the length of the drill chamfer.

    d. 1/2 the length of the drill diameter.

30. What part of the drill press limits the travel of the cutting tool?

    a. Quill

    b. Feed handle

    c. Quill depth stop

    d. Length of the spindle throat

31. The RPM for counterboring and spotfacing should be

    a. 30%–40% of drilling speed.

    b. 20%–30% of drilling speed.

    c. 30%–40% of drilling speed.

    d. 50%–60% of drilling speed.

32. If a 3/8"-16 bolt had a requirement to be 5/8" deep in a workpiece, how many turns would be needed by the tap?

    a. 16

    b. 10

    c. 13

    d. 15

SECTION 5: Turning

UNIT 1 | Introduction to the Lathe

Name: _____ Date: _____

Score: _____ Text pages 352–363

Learning Objectives

After completing this unit, you should be able to:

- Explain the principal operation of a lathe
- Identify and explain the functions of the parts of the lathe
- Explain how lathe size is specified

Carefully read the unit, and then answer the following questions in the space provided.

Fill in the Blank

1. Feed on the lathe is measured by how far the cutting tool _____ each time the spindle turns _____ _____.

2. The top of the lathe bed contains precision rails called _____.

3. _____ feed is the motion of the tool traveling along the ways between the head and tailstock.

4. Tool movement when cross feeding is _____ to the ways.

5. A _____ _____ transmits power to the cross slide and/or the carriage.

6. Over time, the sliding motion of the cross slide and compound rest can cause wear in the dovetail slide. A small wedge-shaped piece known as a _____ can be used as an adjustment to compensate for this wear.

Short Answer

7. Explain the purpose of a lathe feed rack.

_____

_____

8. How does a machinist determine the size of a lathe?

_____

_____

9. How would a machinist use a tailstock?

_____

_____

Name: _____ Date: _____

10. List five major parts of the lathe apron.

_____

_____

_____

_____

_____

11. Which hand wheel would a machinist use to manually turn down a workpiece parallel to the ways?

_____

12. Which hand wheel would a machinist use to manually turn down a workpiece perpendicular to the ways?

_____

True or False

13. The feed rod is a long shaft, either round or hexagonal, that transmits power to the carriage apron during threading.   _____

14. The size of the lathe bed determines the length of the part that can be turned.   _____

15. The cross slide can be adjusted to cut angles on the workpiece.   _____

16. On a geared head lathe, the spindle should be running when changing gears.   _____

17. The half-nut lever is used to cut threads.   _____

Identification Matching

18. Identify the parts of the lathe by writing their names in the blanks in the following diagram.

    a. Feed rack

    b. Leadscrew

    c. Feed rod

    d. Spindle clutch lever

©Cengage Learning 2012

19. Identify the parts of the lathe by writing their names in the blanks on the following diagram.

    a. Apron hand wheel

    b. Feed change knob

    c. Feed reverse knob

    d. Half-nut lever

    e. Feed control clutch

    f. Thread dial

    g. Spindle clutch lever

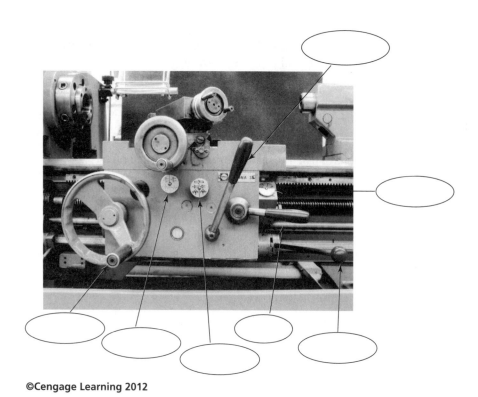

©Cengage Learning 2012

Multiple Choice

20. The spindle of a lathe uses an internal _____ to hold workpiece support accessories.

    a. drawbar

    b. taper

    c. split nut

    d. collet nut

21. In order to change speeds (RPM) on a belt-driven headstock, the operator must

    a. stop the machine and change the lever positions.

    b. leave the machine running and change the lever position.

    c. stop the machine and manually change the belt location.

    d. leave the machine running in low and manually change the belt position.

22. Feed on a lathe is referred to in which of the following units?

    a. SFPM

    b. FPM

    c. IPM

    d. FPR

23. What is the relationship between the sliding motion of the saddle and the location of the bed ways?

    a. Parallel

    b. Perpendicular

    c. Angular

    d. Adjacent

24. A machinist notices that when turning the cross-slide feed handle, there is no immediate movement of the cross slide. This could be caused by which of the following?

    a. No lubrication

    b. Improper tool setup

    c. Gibs adjusted incorrectly

    d. Slide bushing and screw wear

25. On a manual lathe, the operator control panel for tool movement is located on the

    a. saddle.

    b. apron.

    c. carriage.

    d. ways.

26. Bed length is one of the factors that determines the size of a lathe. What is the other factor?

    a. Motor size

    b. Spindle capacity

    c. Swing

    d. Tool capacity

UNIT 2 Workholding and Toolholding Devices for the Lathe

Name: _____ Date: _____

Score: _____ Text pages 364–384

Learning Objectives

After completing this unit, you should be able to:

- Explain the differences between universal-type and independent-type chucks
- Explain the function and application of a three-jaw universal chuck
- Explain the function and application of a four-jaw independent chuck
- Explain the function and application of collets
- Explain the application of various types of lathe centers and related equipment
- Identify and describe the function of mandrels
- Identify and explain the applications of a steady rest and follow rest
- Identify and explain functions of various toolholding devices

Carefully read the unit, and then answer the following questions in the space provided.

Fill in the Blank

1. A _____ is a cylindrical sleeve with several slits. When drawn into a tapered bore, it contracts and grips the workpiece.
2. Collets are made of both _____ and _____ material.
3. A lathe center is a cylindrical steel device with a _____ -degree, _____ -angled point on one end and a Morse taper on the other end.
4. If a tool or toolholder is extended too far, _____ could occur.
5. Drill chucks with _____ shanks can be mounted in the tailstock to perform holemaking operations using straight-shank tools.
6. If a machinist is to turn down the complete outside diameter of workpiece that contains a hole completely through the center, the machinist could use a _____ to hold workpiece.

Short Answer

7. List the parts of a rocker tool post.

_____

_____

8. Name the three main styles of tool posts.

_____

_____

_____

9. Name the taper type commonly found in lathe tailstocks.

_____

_____

10. What types of workpieces might be good candidates for mounting to a faceplate?

_____

_____

11. Explain the easiest and quickest way to check alignment of a tailstock.

_____

_____

12. What accessories can be used to support long workpieces during turning?

_____

_____

13. Briefly describe the difference between a quick-change toolholder and an indexable tool post.

_____

_____

True or False

14. A live center and dead center have the same function in a tailstock. _____

15. A drive plate is mounted to a drive dog. _____

16. Standard engine lathes can machine holes perpendicular to the workpiece. _____

17. Large work is commonly secured to faceplates, which can cause workpiece imbalance. _____

18. One of the most precise spindle workholding devices is the four-jaw chuck. _____

19. The jaws on a three-jaw chuck move independently. _____

20. A follower rest is bolted to the carriage and travels along the part with the cutting tool. _____

Multiple Choice

21. The main difference between a three-jaw chuck and a four jaw-chuck is the
    a. size of material they can hold.
    b. physical size of the chuck.
    c. jaw tightening screw.
    d. movement of the jaw.

22. Which of the following chucks provides the best accuracy for runout?
    a. University chuck
    b. Four-jaw chuck
    c. Three-jaw chuck
    d. Soft-jaw chuck

23. The most accurate way to support a workpiece from a spindle is by using a
   a. four-jaw chuck.
   b. three-jaw chuck.
   c. collet system.
   d. universal chuck.

24. A typical lathe center has an included angle of
   a. 82 degrees.
   b. 90 degrees.
   c. 30 degrees.
   d. 60 degrees.

25. The main advantage of supporting work and turning between centers is
   a. the work can be removed without losing accuracy.
   b. tool clearance is improved and almost eliminated.
   c. tool set up is easier.
   d. None of the above

26. The main difference between a steady rest and a follower rest is the
   a. third point of contact provided.
   b. position of the rest on the lathe.
   c. length and position of the support provided.
   d. All of the above
   e. None of the above

27. The oldest and simplest toolholder used with a lathe is the
   a. quick-change toolholder.
   b. rocket-type toolholder.
   c. indexable toolholder.
   d. square-box toolholder.

28. A special precision cylindrical shaft that can be inserted through a centerbore of the workpiece is called a
   a. drawbar.
   b. draw tube.
   c. mandrel.
   d. center shaft.

UNIT 3	Machining Operations on the Lathe

Name: _____ Date: _____

Score: _____ Text pages 385–414

Learning Objectives

After completing this unit, you should be able to:

- Explain the relationship between depth of cut and diameter
- Compare and contrast roughing and finishing operations
- Explain lathe speed and feed terms, and calculate spindle speeds and machining time
- Identify basic cutting tools and cutting-tool geometry
- Explain lathe safety precautions
- Describe and explain the purpose of facing, turning, and shouldering operations
- Describe and explain lathe holemaking operations
- Explain how to use taps and dies to cut threads on the lathe
- Define and explain form cutting
- Describe and explain grooving and cutoff operations
- Describe and explain the purpose and process of knurling

Carefully read the unit, and then answer the following questions in the space provided.

Fill in the Blank

1. Facing is used to machine the _____ of the workpiece.
2. Turning is used to reduce the _____ of the workpiece.
3. Never use a file without a _____.
4. Never use _____ _____ to clean the lathe.
5. Never make machine adjustments, take measurements, or clean a lathe when the machine is
   _____.
6. Cutoff operations should never be performed when work is held _____
   _____.
7. When grooving or parting, the cutting tool should be set at a _____ angle to the ways.
8. When knurling, use a _____ spindle speed and a _____ feed rate.

Short Answer

9. List 10 safety rules for a lathe. Which is more important?

_____

_____

_____

_____

_____

_____

_____

_____

_____

_____

_____

10. Briefly define *boring*.

_____

_____

11. List two reasons knurling is performed.

_____

_____

12. Explain the importance of cutting-tool side rake.

_____

_____

True or False

13. Roughing operations use lower cutting speeds than finishing operations. _____

14. Finishing operations use higher feed rates than roughing operations. _____

15. Feed rate movements are the same for perpendicular and parallel cuts on a workpiece. _____

16. As the workpiece becomes smaller during turning, the RPM decreases. _____

17. Lathe and drill presses use the same nomenclature/unit of measure for feed. _____

18. When filing, the machinist should stand to the right of the lathe headstock. _____

Identification Matching

19. Sketch the basic geometry of a lathe tool. Specifically the back and side rakes.

20. Identify and label the proper turning operations in the given pictures.

    a. Boring

    b. Grooving

    c. Knurling

    d. Facing

    e. Reaming

    f. Turning

    g. Drilling

_____

©Cengage Learning 2012

_____

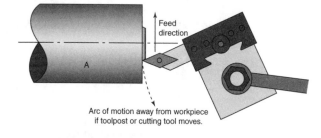

Feed direction

A

Arc of motion away from workpiece
if toolpost or cutting tool moves.

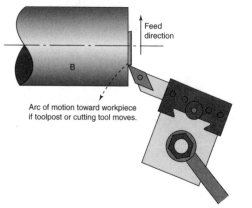

Feed direction

B

Arc of motion toward workpiece
if toolpost or cutting tool moves.

©Cengage Learning 2012

_____

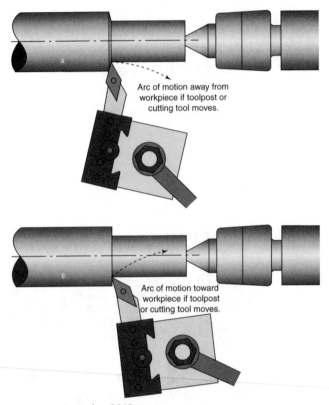

Arc of motion away from workpiece if toolpost or cutting tool moves.

Arc of motion toward workpiece if toolpost or cutting tool moves.

©Cengage Learning 2012

_____

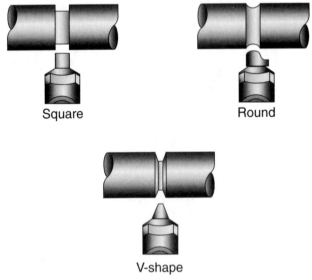

Square   Round

V-shape

©Cengage Learning 2012

_____

©Cengage Learning 2012

_____

©Cengage Learning 2012

_____

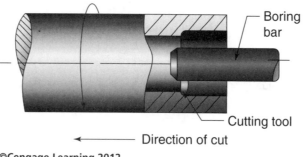

Boring bar

Cutting tool

Direction of cut

©Cengage Learning 2012

Computations

21. Determine the proper RPM for a 3" piece of cold-rolled steel with a cutting speed of 90 SFPM.

_____

22. Calculate the machining time for a part 25" in length of 10"-diameter AISI/SAE 4140 steel using a cutting speed of 110 SFPM and a feed rate of .008 IPR.

_____

Multiple Choice

23. If a feed rate was .015 IPR", what would this mean?

    a. The tool would advance .015" during each pass

    b. The tool would advance .015" in depth during each pass

    c. The tool would advance .015" per each revolution of the spindle

    d. The tool would advance .015" per each in-feed pass

24. The three main factors that determine machining time calculations are

    a. material, hardness of material, and tooling.

    b. tooling, machine, and lubricant.

    c. length of part, RPM, and feed.

    d. feed, cutting-tool shape, and hardness of the material.

25. The cutting direction of a left-handed cutting tool is

    a. right to left.

    b. left to right.

    c. either direction.

    d. None of the above

26. When setting up a cutting tool in relationship to the workpiece, the cutting tool must have the

    a. tip touching the workpiece.

    b. center of the work aligned with the top of the tool.

    c. center of the work aligned with the center of the tool.

    d. cutting edge angle touching the workpiece.

27. A machinist is turning a piece of material down from 5" to 2" in diameter. The cutting speed is 90 SFPM. What is the difference in RPM from 5" to 2"?

    a. 85 RPM

    b. 103 RPM

    c. 250 RPM

    d. 90 RPM

28. When filing on a lathe, which hand should be the primary holder of the file?

    a. Right only

    b. Left only

    c. Both right and left, equally spaced on handle

    d. Left on handle, right supporting motion

SECTION 5: Turning

UNIT 4	Manual Lathe Threading

Name: _____   Date: _____

Score: _____   Text pages 415–433

Learning Objectives

After completing this unit, you should be able to:

- Identify the parts of a thread and define thread terminology
- Describe the difference between left-hand and right-hand threads
- Identify and describe the different classes of fit
- Accurately locate thread reference data from a chart
- Perform calculations required for thread cutting
- Describe the proper setup of a workpiece and cutting tool for threading
- Describe the process of lathe thread cutting
- Describe various methods of thread measurement

Carefully read the unit, and then answer the following questions in the space provided.

Fill in the Blank

1. _____ is the distance a thread will move in relationship to its mating thread in one revolution.

2. In the Unified National thread system, external threads are indicated by the letter _____ , while internal threads are indicated by the letter _____ .

3. When cutting threads on the lathe tool, in-feed will be controlled by the _____ , _____ which should be positioned at _____ degrees for 60-degree V-threads.

4. The _____ _____ , also known as the fishtail gage, is used to align the tip of the threading tool perpendicular to the workpiece.

5. _____ _____ threads are often used in low-pressure couplings where the use of pipe compound or pipe tape is acceptable.

Short Answer

6. What type of thread has an asymmetrical form, and what is its purpose?

_____

_____

7. What will result when the diameter to be threaded has runout when mounted in the machine?

_____

_____

8. After making the initial pass when cutting threads, it is a good idea to double check the pitch of the thread with what device?

_____

_____

9. What is TPI?

_____

_____

10. Name three tools to measure threads. Which is the most accurate for measuring?

_____

_____

_____

11. Why are acme threads used for power transmission?

_____

_____

True or False

12. A triple-lead thread will advance faster than a one-lead thread when rotated. _____

13. A 3B thread will be more expensive to produce than a 1B thread. _____

14. The compound feed rate determines the angle of the thread. _____

15. The pitch diameter is the key factor for a good-fitting thread. _____

Identification Matching

16. Using the appropriate letter, label the parts of the Unified thread shown in the sketch.

    a. Major diameter

    b. Minor diameter

    c. Pitch diameter

    d. Crest

    e. Root

    f. Flank

    g. Thread angle

    h. Helix angle

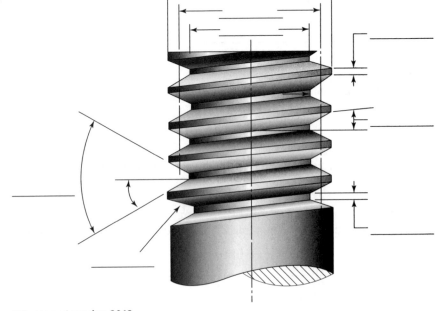

Computations

17. If the quick-change gear box is set to the whole number 13, how many threads per inch will be produced on a 1-inch-diameter shaft?

_____

18. What are the upper and lower limits for the measurement over wires of a 7/8-14 UNF 3A thread? Use .0500 wires for the calculation.

    a. upper: _____

    b. lower: _____

19. Refer to the chart in Figure 5.4.6 in the Unit to determine the basic, maximum, and minimum limit of the major diameter of 3/8-16 2A UNC.

_____

20. What is the estimated compound rest in-feed depth on a 1/4-28 UNF thread?

_____

Multiple Choice

21. The example of a micrometer's movement of .025" per one revolution is referred to as

    a. helix angle movement.

    b. lead of the thread.

    c. pitch of the thread.

    d. thread per inch.

22. If a bolt had three leads, it would

    a. advance a nut faster than a single-lead bolt.

    b. advance a nut at the same speed as a single-lead bolt.

    c. not advance a nut; a three-lead bolt does not exit.

    d. not make a difference either way.

23. Which of the following characters reflects an internal or external thread in the following specification: 11/16"-16 UNC 3A?

    a. A

    b. U

    c. N

    d. C

24. The main factor of a thread that determines the class of fit is the

    a. hex angle.

    b. root size.

    c. pitch diameter.

    d. lead of the thread.

25. When threading, the compound rest is set up for depth of in-feed at an angle of

    a. 29 1/2 degrees.

    b. 60 degrees.

    c. 30 degrees.

    d. None of the above

26. The control on the lathe that sets the threads per inch of feed is called the

    a. headstock.

    b. feed control level.

    c. quick-change gear box.

    d. feed bar.

UNIT 5	Taper Turning

Name: _____   Date: _____

Score: _____   Text pages 434–447

Learning Objectives

After completing this unit, you should be able to:

- Define a taper
- Demonstrate understanding of taper specification methods
- Perform taper calculations
- List methods of turning tapers and their benefits and drawbacks
- Demonstrate understanding of setup procedures for taper turning methods

Carefully read the unit, and then answer the following questions in the space provided.

Fill in the Blank

1. A 3/4 taper means that in 1" of length, the diameter would change by _____ of an inch.
2. When cutting a taper, to calculate the spindle-speed or RPM, the machinist would use the _____ diameter of the workpiece.
3. If the compound rest feed handle is too tight to turn, the _____ can be adjusted to free movement.
4. Excessive play in the lead screw of the carriage and compound rest when reversing directions is known as _____.
5. A method to determine if an angle is correct on set up is to use a _____ , _____.

Short Answer

6. List three drawbacks of the offset tailstock method of cutting tapers.

   _____

   _____

   _____

7. What does TPF stand for?

   _____

   _____

8. What error might result when using the compound-rest taper-turning method?

   _____

   _____

9. What is the common factor when converting TPI to TPF?

   _____

   _____

True or False

10. A tool bit can be positioned to cut tapers with their cutting-edge form. _____

11. TPF and TPI result in the same angle. _____

12. A taper angle is measured from one side of the workpiece center line. _____

13. Taper attachments can perform internal and external operations. _____

14. Compound rests have power feed for cutting tapers. _____

Identification Matching

15. Identify the proper method to cut the given taper on a workpiece.

a. Offset tailstock	_____	**1.** Taper 10" long × .1875"
b. Tool bit formed	_____	**2.** Taper 1" long × .125 TPI
c. Taper attachment	_____	**3.** Chamfer 1/16" × 1/16"
d. Compound rest	_____	**4.** Taper 4" long × .0625 TPI
e. File	_____	**5.** Chamfer 1/8" × 1/8"

Computations

16. If a bar is 2.00" diameter on one end and .75" on the other end, and is 5" long, what is the angle of the taper per inch?

_____

_____

17. A print specifies a 3-degree included angle. What is the corresponding TPI?

_____

_____

18. Calculate the set over of the tailstock given a 15" long workpiece and a .0625" TPI. Is this a safe set up?

_____

_____

Multiple Choice

19. Tapers are identified by two different means in the machining industry. The first is angular dimension, and the second is

    a. total length of part.

    b. diameter of part at one end.

    c. both length of part and diameter change.

    d. None of the above

20. Which of the following lathe set ups would produce a taper about 1" long and 3 degrees on a workpiece?

    a. Compound rest angled

    b. Tailstock offset

    c. Taper attachment

    d. All of the above

    e. None of the above

21. The major advantage of a taper attachment is which of the following?

    a. No backlash during in-feed

    b. Quicker setup time

    c. Power feed can be used

    d. Any length of taper can be machined

22. The limitation of the tailstock offset taper set up is which of the following?

    a. The length of the part to be machined

    b. The degree of the angle to be cut

    c. The size of material that can be machined

    d. None of these are limitations

SECTION 6: Milling

UNIT 1 | Introduction to the Vertical Milling Machine

Name: _____ Date: _____

Score: _____ Text pages 448–463

Learning Objectives

After completing this unit, you should be able to:

- Identify the components of the vertical milling machine
- Explain the function of the components of the vertical milling machine

Carefully read the unit, and then answer the following questions in the space provided.

Fill in the Blank

1. The _____ _____ passes through the spindle and is used to secure toolholding devices into the machine spindle.
2. The _____ _____ can be used to quickly stop spindle rotation or to lock the spindle to prevent rotation.
3. The X-axis moves to the _____ and _____ from the operator.
4. The Y-axis moves _____ and _____ from the operator.
5. A _____ _____ locks the elevating crank in place after positioning the knee.
6. The standard taper for a vertical mill spindle is an _____ .

Short Answer

7. Define the term *DRO* and explain how it is used.

   _____

   _____

8. Machinists' maintenance responsibilities should include lubricating the vertical milling machine using what type of system? How frequently should this be done?

   _____

   _____

9. Explain the function of a drawbar.

   _____

   _____

10. What part of the vertical mill is used to control quill feed depth?

    _____

    _____

11. Explain the difference between a variable-speed head and step-pulley head.

_____

_____

True or False

12. Angles can be cut on a vertical milling machine by titling the head. _____

13. An accurate way to measure a titled head on a vertical machine is by using a protractor. _____

14. Power quill feed is adjustable from a range of 0.0015" to .006". _____

15. The table can be moved up or down, manually or by power. _____

Identification Matching

16. Using the appropriate letter, label the parts of the vertical milling machine in the photo/illustration.

    a. Knee

    b. Saddle

    c. Turret

    d. Head

    e. Table locks

    f. Base

    g. Ram

    h. Table crank

    i. Table

    j. Saddle crank

    k. Knee crank lock

    l. Saddle lock

  m. Column

    n. Knee elevating crank

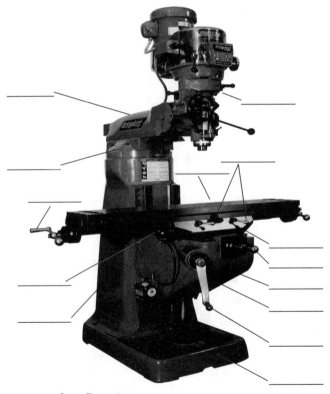

Courtesy of Hardinge, Inc.

17. Label the parts of the vertical milling machine head in the photo/illustration.

    a. Quill feed selector knob

    b. Micrometer adjusting knob

    c. High/low range switch

    d. Feed control lever

    e. Variable speed dial

    f. Speed change handwheel

    g. Spindle

    h. Quill

    i. Spindle brake

    j. Power feed transmission engagement crank

    k. Quill stop

    l. High/neutral/low lever

    m. Manual feed handwheel

    n. Feed reverse knob

    o. Quill lock

    p. Quill feed handle

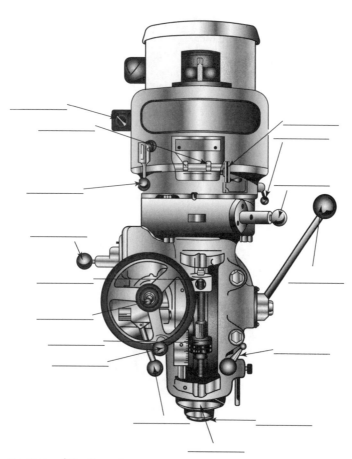

Courtesy of Hardinge, Inc.

Multiple Choice

18. If the quill is locked, which part of the mill can be moved to adjust the Z axis?

    a. Head

    b. Column

    c. Knee

    d. Table

19. The taper that is most commonly found in a manual vertical mill spindle is called a

    a. MT40.

    b. CAT8.

    c. R-8.

    d. MS40.

20. When adjusting the gear range from high to low on a geared head vertical mill, it is important to

    a. have the motor running.

    b. not have the motor running.

    c. have the spindle brake on at all times.

    d. have no tools in the spindle.

21. The precision of a vertical mill is the relationship between the

    a. center of the spindle and the plane of the table.

    b. center of the motor and table ways.

    c. center of the knee and column.

    d. center of the ram and column.

SECTION 6: Milling

<table>
<tr><td>UNIT 2</td><td>Tools, Toolholding, and Workholding for the Vertical Milling Machine</td></tr>
</table>

Name: _____ Date: _____

Score: _____ Text pages 464–480

Learning Objectives

After completing this unit, you should be able to:

- Identify and explain the use of various cutting tools used on the milling machine
- Identify and explain the use of various toolholding devices used on the milling machine
- Identify and explain the use of various workholding devices used on the milling machine
- Identify which type of holding device should be used in different situations
- Identify and explain applications of a mill vise
- Identify and explain applications of hold-down clamps

Carefully read the unit, and then answer the following questions in the space provided.

Fill in the Blank

1. Some endmills come with a special flat on the shank so that they may be secured with a set screw. This is called a _____ .

2. Mill vises may be mounted atop a(n) _____, which allows the vise to be rotated and secured in the desired position.

3. When using hold-down clamps, chose the _____ clamp strap that will allow the desired machining operation to be performed both efficiently and safely.

4. Slitting saws and form cutters have a straight hole through their center with a single keyway and are mounted to the spindle with a _____ .

5. End cutting endmills can also be classified as either _____ or _____ -cutting.

6. The most important step before any milling operation starts is to check the workpiece _____ and make sure the cutter is mounted securely.

Short Answer

7. List the three main parts of the inserted milling cutter.

_____

_____

_____

8. List two workholding applications where a milling vise would be best.

_____

_____

9. Why does the workpiece holding device determine accuracy and time of machining task?

_____

_____

10. Briefly describe the procedure when securing an endmill in a collet.

_____

_____

11. Explain the difference between a face mill and shell endmill.

_____

_____

True or False

12. Vacuum plates are used during heavy milling operations.   _____

13. Mill cutting tools are fragile and damage easily.   _____

14. HSS endmills can be reground to their original shape and size.   _____

15. To mill an internal radius, a concave-type endmill can be used.   _____

16. Fly cutters should not be used for taking heavy roughing passes.   _____

17. Many drill press accessories can be used on a vertical mill.   _____

Identification/Matching

18. Circle the endmill(s) pictured below that has/have center-cutting capabilities.

©Cengage Learning 2012

19. Match the best workpiece holding device to the given workpiece.

a. Special fixture	_____	**1.** Oddly shaped part 8" × 4" × 16"
b. Magnetic plate	_____	**2.** Small casting with bevels
c. Vacuum plate	_____	**3.** Aluminum bar 1/2" × 2" × 4"
d. Adhesive plate	_____	**4.** Thin sheet metal
e. Collet fixture	_____	**5.** Large casting with odd shapes
f. Fixed mill vise	_____	**6.** Production run of soft material
g. Angled vise	_____	**7.** Iron plate with welded stand-offs
h. Toe clamp	_____	**8.** HRS 1/2" × 6" × 24"
i. Toggle clamp	_____	**9.** Decorative trim
j. Angle plate	_____	**10.** CRS bar 1/2" diameter

Multiple Choice

20. Milling cutters and endmills are very brittle and should be stored
   a. to avoid contact with each.
   b. in plastic cases.
   c. without contact to the cutting surface.
   d. All the above
   e. None of the above

21. The main difference between a roughing endmill and finishing endmill is the
   a. length of the cutting edge.
   b. shape of the cutting edge.
   c. size of the endmill.
   d. type of shank.

22. The best endmill for soft, gummy aluminum machining would be a
   a. four-flute.
   b. two-flute.
   c. six-flute.
   d. ball mill.

23. A fly cutter uses how many cutting-tool edges?
   a. One
   b. Two
   c. Three
   d. Four

24. If an internal fillet is needed on a workpiece, what would be the best cutter to use?
   a. Standard endmill
   b. Concave cutter
   c. Convex cutter
   d. Radius cutter

25. The mill vise can be mounted to the table
    a. parallel to the X-axis.
    b. perpendicular to the Y-axis.
    c. offset to the spindle axis.
    d. All of the above
    e. None of the above

26. A workholding device that is custom-made for a specific workpiece shape is called a
    a. jig.
    b. fixture.
    c. soft table jaw.
    d. custom clamp.

SECTION 6: Milling

<table>
<tr><td>UNIT 3</td><td>

Vertical Milling Machine Operations
</td></tr>
</table>

Name: _____   Date: _____

Score: _____   Text pages 481–519

Learning Objectives

After completing this unit, you should be able to:

- Describe vertical milling machine safety practices
- Describe the purpose and process of tramming the milling machine head
- Calculate speeds and feeds for milling operations
- Explain how to use an edge finder to establish a reference location
- Explain how to use an indicator to locate the center of a part feature
- Explain the process of boring on the milling machine
- Define conventional and climb milling
- Explain the process of squaring a block on the milling machine
- Explain the basic steps of milling rectangular pockets

Carefully read the unit, and then answer the following questions in the space provided.

Fill in the Blank

1. A process called _____ sets the perpendicularity of the head to the table.
2. An accurate way to align a workpiece to an axis is to use a _____
   _____.
3. When roughing, use slightly _____ speeds, _____ feed rates,
   and _____ cuts.
4. The _____ of an endmill determines the radius of a corner.
5. The bigger the cutter, the _____ the spindle speed.
6. The softer the material, the _____ the spindle speed.
7. Squaring a workpiece means machining the sides of a workpiece _____ and
   _____ to each other.

Short Answer

8. Define and briefly describe the process of aligning the spindle so it is perpendicular to the milling machine table.

   _____

   _____

9. What is the formula for calculating IPM for milling operations?

_____

_____

10. How much finish material should be left on the side after a roughing cut?

_____

_____

11. Briefly describe the steps to use an edge finder to locate the edge of a workpiece.

_____

_____

12. What tool is used for boring on the vertical mill that allows a boring bar to be offset to adjust the size of the hole being machined?

_____

_____

13. Sketch the first four steps of the block squaring process.

14. Briefly describe two methods for positioning an endmill in the center of a shaft.

_____

_____

15. What is a coordinate map?

_____

_____

True or False

16. Leaving measuring tools on the mill table is a safe practice. _____

17. The mill brake is used when securing a cutter. _____

18. Feed per tooth is also called "chip load." _____

19. When roughing out a pocket, use the largest-diameter endmill you can find. _____

20. Climb milling pulls the workpiece into the cutter. _____

21. The protractors found on the machine and table vise are very accurate. _____

Computations

22. Calculate RPM and feed rate in IPM for a 3/4"-diameter two-flute HSS endmill cutting aluminum at 250 SFPM with a 0.004" FPT.

_____

_____

23. Calculate the RPM for a boring bar: material CRS, cs = 90, diameter of hole 2".

_____

_____

Multiple Choice

24. The process that is used to adjust the spindle perpendicular to the top surface of the table is called
   a. trimming.
   b. positioning.
   c. tramming.
   d. indicating.

25. The three factors for determining milling feed rate in IPM are
   a. material hardness, tooling tool type, and lubrication.
   b. RPM, toolholder type, and material hardness.
   c. lubrication, tool material, and workholding device.
   d. RPM, FPT, and number of teeth.

26. The feed handle on the table is incremented in how many units?
   a. .100"
   b. .150"
   c. .200"
   d. .250"

27. The average speed for an edge finder is
   a. 100 RPM.
   b. 1,100 RPM.
   c. 500 RPM.
   d. 200 RPM.

28. If an edge finder with a diameter of .200" is used to find a center, how much offset would need to be adjusted to bring it to center?
   a. .200"
   b. .400"
   c. .050"
   d. .100"

29. The main difference between conventional milling and climb milling is the
   a. direction of feed.
   b. size of the cutter.
   c. depth of cut.
   d. rate of feed.

UNIT 4 Indexing and Rotary Table Operations

Name: _____ Date: _____

Score: _____ Text pages 520–529

Learning Objectives

After completing this unit, you should be able to:

- Identify what processes can be performed on a rotary table or dividing head
- Identify the basic parts of a rotary table and dividing head
- Understand the basic setup and operation of the rotary table and dividing head
- Perform direct and simple indexing calculations

Carefully read the unit, and then answer the following questions in the space provided.

Fill in the Blank

1. Two types of indexing are possible with the indexing head. _____ requires a plate with multiple hole patterns that are used to position the crank handle. _____ is used to perform rapid indexing for only a few division combinations.

2. The number of divisions is limited by the number of _____ or _____ in the direct index plate.

3. An important aspect of the rotary table is the alignment of the _____ _____ of the table.

4. _____ turns of the crank results in _____ turn of the spindle on simple indexing.

5. It is good practice to check the _____ of the centers while the test bar is mounted because some tailstocks have a height adjustment.

6. A whole number represents _____ complete turns of the hand crank.

Short Answer

7. List three uses of a rotary table.

   _____

   _____

   _____

8. What is the main function of an indexing head?

   _____

   _____

9. List the parts of an indexing head.

_____

_____

10. What is the most important step in aligning an indexing head?

_____

_____

11. What is the difference between simple and direct indexing?

_____

_____

True or False

12. Over-travel of a cutter on a tangent cut will produce an undercut.   _____

13. An indexing head can be manual or power driven.   _____

14. If an even number of turns is required, care must be used to select the proper indexing plate.   _____

15. Collets are used in all indexing heads to hold the workpiece.   _____

16. The current hole is counted when using indexing plates.   _____

Computations

17. If the indexing head gear train is going to be used to perform simple indexing, calculate the number of whole and partial turns using the index plates below for the following situations:

Plate #1 has hole circles with 15, 16, 17, 18, 19, and 20 holes

Plate #2 has hole circles with 21, 23, 27, 29, 31, and 33 holes

Plate #3 has hole circles of 37, 39, 41, 43, 47, and 49 holes

   a. Make 32 divisions

   _____ whole turns and _____ holes in a _____ -hole circle

   b. Make 39 divisions

   _____ whole turns and _____ holes in a _____ -hole circle

   c. Make 56 divisions

   _____ whole turns and _____ holes in a _____ -hole circle

Multiple Choice

18. Rotary tables are aligned to the center of the spindle with a

   a. dial test indicator.

   b. dial plunge indicator.

   c. dial depth gage.

   d. All of the above

   e. None of the above

19. The typical workholding device used on a rotary table is(are) the

   a. three-jaw chuck.

   b. spring clamp.

   c. toe clamps.

   d. faceplates.

20. The function of a rotary table is to

    a. mill straight lines on large castings.

    b. move the cutter to make arcs and circles.

    c. move the workpiece to make arcs and circles.

    d. mill linear lines on small-diameter workpieces.

21. When an indexing head uses its spindle flange to determine the proper division needed, it is called

    a. simple indexing.

    b. direct indexing.

    c. complex indexing.

    d. spindle indexing.

SECTION 7: Grinding

<table>
<tr><td>UNIT 1</td><td>Introduction to Precision Grinding Machines</td></tr>
</table>

Name: _____   Date: _____

Score: _____   Text pages 530–538

Learning Objectives

After completing this unit, you should be able to:

- Explain the benefits of precision grinding
- Identify various types of grinders and their capabilities
- Identify the parts of a surface grinder

Carefully read the unit, and then answer the following questions in the space provided.

Fill in the Blank

1. Grinding to tolerances of _____ is not unusual when operating a precision grinder.
2. Surface finishes as fine as _____ microinches are common when grinding.
3. Another name for a vertical spindle rotary surface grinder is a _____.
4. The cross-feed hand wheel is rotated to provide movement to the _____-axis.
5. The _____ surface grinder is probably the most common type of precision grinding machine used in the machining industry.

Short Answer

6. What are the two surface grinder table movements that are similar to movements of the vertical milling machine?

   _____

   _____

7. Why can cross feed be set to reverse direction?

   _____

   _____

8. List three ways to mount a workpiece for cylindrical grinding.

   _____

   _____

   _____

9. What is the main function of a jig grinder?

   _____

   _____

True or False

10. The main difference between off-hand grinding and precision grinding is the amount of material removed.   _____

11. After setting the amount of travel for both axes, the cross feed will continuously cycle back and forth under the wheel.   _____

12. Vertical spindle grinders can use either reciprocating or rotary table motion.   _____

13. A tool grinder is considered an off-hand grinder.   _____

14. Surface grinders can remove both ferrous and non-ferrous materials.   _____

Identification Matching

15. Match the given precision grinding machine to the grinding task.

a. Horizontal spindle surface grinders   _____   **1.** Uses a solid cylindrical or cup-shaped wheel

b. Vertical spindle surface grinders   _____   **2.** Work is rotated against the rotation of the grinding wheel

c. Cylindrical grinders   _____   **3.** Uses a spindle-mounted grinding wheel

d. Centerless grinders   _____   **4.** Used to create customized tools from HSS or carbide tool blanks

e. Tool grinders   _____   **5.** Uses the periphery of a solid wheel to grind workpiece surfaces

f. Jig grinders   _____   **6.** Work is supported by a work rest while being fed between the grinding wheel and a regulating wheel

Multiple Choice

16. The primary purpose of a surface grinder is to produce
   a. flat surfaces.
   b. curved surfaces.
   c. angled surfaces.
   d. All of the above
   e. None of the above

17. Precision grinding can produce a surface finish as fine as
   a. a 200-grit finish.
   b. 16 microinches.
   c. 50 cu.
   d. 100.

18. A typical horizontal spindle surface grinder uses the _____ of the grinding wheel to remove material.
   a. edge
   b. side
   c. periphery
   d. ID

19. A grinder that performs machining tasks similar to that of a lathe is called a
   a. horizontal grinder.
   b. vertical grinder.
   c. jig grinder.
   d. cylindrical grinder.

20. Which of the following is NOT a characteristic of a jig grinder?
    a. Has precise movement of the X- and Y-axes
    b. Uses a spindle-mounted grinding wheel
    c. Used for sharpening of machine tools
    d. Produces a smooth finish, accurate hole

UNIT 2	Grinding Wheels for Precision Grinding

Name: _____   Date: _____

Score: _____   Text pages 539–544

Learning Objectives

After completing this unit, you should be able to:

- Identify grinding wheel shapes
- Explain the grinding wheel identification system
- Define the types of abrasives used to make grinding wheels
- Explain grain size
- Explain the hardness scale of grinding wheels
- Explain the different types of grinding-wheel bonds
- Explain wheel structure
- Describe grinding-wheel characteristics suitable for various applications
- Describe the use of superabrasives for precision grinding

Carefully read the unit, and then answer the following questions in the space provided.

Fill in the Blank

1. Grain sizes of grinding wheels range from _____ to _____.
2. Wheel grades are identified from _____ to _____ by the letters _____ through _____.
3. The _____ bond is the most common bonding agent used in grinding wheels.
4. The letter _____ is used for diamond wheels.
5. Grinding wheels that are used to grind extremely hard materials are called _____.
6. The ability of the individual abrasive grains to fracture during grinding to create new, sharp cutting edges is called _____.

Short Answer

7. Explain the advantages of an open-structured grinding wheel.

_____

_____

8. What are the advantages and disadvantages of the friability of grinding wheels?

_____

_____

9. What is the disadvantage in using superabrasives?

_____

_____

10. What two types of silicon carbide grades are available, and what material do they grind?

_____

_____

True or False

11. Aluminum oxide defines the structure of the grinding wheel.   _____

12. The letter E identifies shellac wheels.   _____

13. Cubic boron nitride wheels are harder than diamond wheels.   _____

14. Surface grinding can be performed on most all materials.   _____

Identification Matching

15. Match the correct bond type to the given definition.

    a. Vitrified   _____   **1.** Is a natural or organic bond; it can deteriorate over time and become brittle and weak

    b. Silicate   _____   **2.** Used to make very thin wheels

    c. Rubber   _____   **3.** Used for rough off-hand grinding under harsh conditions

    d. Resinoid   _____   **4.** Are hard but too brittle to withstand heavy pressure or shock

    e. Shellac   _____   **5.** Used to grind very thin parts and edges of hard materials

Multiple Choice

16. The best type of grinding wheel for tool and cutter grinding is a
    a. cup wheel.
    b. saucer wheel.
    c. dish-shaped wheel.
    d. All the above
    e. None of the above

17. Which character defines the type of wheel material? 38 A 60 J 8 V BE
    a. A
    b. J
    c. V
    d. BE

18. Which character defines the grain structure? 38 A 60 J 8 V BE
    a. 38
    b. A
    c. J
    d. 8

19. Silicon carbide grinding wheels are designated by the letter
- **a.** A.
- **b.** B.
- **c.** C.
- **d.** S.

20. What is the grit of the wheel? 38 A 60 J 8 V BE
- **a.** 38
- **b.** 60
- **c.** 8
- **d.** BE

21. The spacing between the individual grains of a grinding wheel is called
- **a.** bond.
- **b.** grain size.
- **c.** structure.
- **d.** grade.

22. Grinding wheels that are used to grind extremely hard materials are called
- **a.** diamond abrasive.
- **b.** CBN.
- **c.** superabrasives.
- **d.** All of the above
- **e.** None of the above

SECTION 7: Grinding

UNIT 3 | Surface Grinding Operations

Name: _____ Date: _____

Score: _____ Text pages 545–561

Learning Objectives

After completing this unit, you should be able to:

- Describe surface grinder safety guidelines
- Explain the basic process of mounting and dressing surface grinder wheels
- Identify and explain the use of common workholding devices used for surface grinding
- Explain the process of grinding parallel, perpendicular, and angular surfaces
- Describe methods for side grinding of vertical surfaces
- Describe common grinding problems and solutions

Carefully read the unit, and then answer the following questions in the space provided.

Fill in the Blank

1. When using a surface grinder with coolant, the main cause of scratches on a workpiece is usually
   _____ _____.

2. The limit for side grinding is usually _____ of an inch.

3. Generally, the surface to be ground is positioned _____ to the magnetic chuck.

4. A brake truing device contains a _____ _____ wheel mounted on a spindle with an automatic braking system.

5. Aluminum oxide and silicon carbide wheels are dressed and trued simultaneously using a _____ or _____ dresser.

6. Never mount a wheel with an RPM rating that is _____ than the spindle RPM.

Short Answer

7. List the steps to be performed before any surface grinding is started.

   _____

   _____

8. If the workpiece is leaving a burn on the surface after a grind, what should be done?

   _____

   _____

9. State three common grinding problems and how to remedy them.

   _____

   _____

   _____

10. List four holding devices that can produce grinded angles.

_____

_____

_____

_____

11. What are you checking for when performing a ring test?

_____

_____

True or False

12. Dressing a grinding wheel concerns the periphery of the wheel. _____

13. A wavy finish could be caused by inadequate truing. _____

14. If the magnetic chuck is worn, grinding is an alternative to fixing the problem. _____

15. A permanent magnet is activated by flipping a lever to energize the magnet. _____

16. Always shut off the wheel and let it come to a complete stop before making workpiece adjustments or taking measurements. _____

17. The speed and feed of dressing the grinding wheel will determine the heat generated during the grinding operation. _____

Multiple Choice

18. The cushion between the grinding wheel and spindle flange is called a

    a. spacer.
    b. blotter.
    c. shim.
    d. None of the above

19. A magnetic chuck can be used on

    a. non-ferrous material.
    b. ferrous material.
    c. austenitic steel.
    d. cast iron.

20. The technique used to secure a workpiece that is taller than its length or width is called

    a. cramping.
    b. dogging.
    c. blocking.
    d. chucking.

21. When setting up a single-point diamond dresser, the dresser should be on which side of the center of the grinding wheel?

    a. Right side
    b. Left side
    c. Neutral
    d. Does not matter which side

22. Which of the following is NOT considered a cause of surface scratches?

    a. No coolant

    b. Dirty coolant

    c. Too much coolant

    d. Loaded grinding wheel

SECTION 8: Computer Numerical Control

UNIT 1 | CNC Basics

Name: _____ Date: _____

Score: _____ Text pages 562–577

Learning Objectives

After completing this unit, you should be able to:

- Identify and describe basic CNC motion-control hardware
- Describe the Cartesian coordinate system
- Describe the polar coordinate system
- Describe the absolute and incremental positioning systems
- Describe the purpose of G- and M-codes
- Describe word addresses
- Describe modal codes
- Describe what a "block" is in CNC programming
- Describe machine motion types
- Describe the main components of a CNC program

Carefully read the unit, and then answer the following questions in the space provided.

Fill in the Blank

1. Write the proper code next to each description:
   a. Linear Interpolation _____
   b. Circular Interpolation—Clockwise _____
   c. Circular Interpolation—Counterclockwise _____
   d. Rapid Traverse _____
   e. Program End _____
   f. Coolant On _____
   g. Coolant Off _____
   h. Tool Change _____

2. CNC programs are written and stored in the _____ _____
   _____ or _____.

3. Linear guides have pressurized _____ systems, and contain _____
   ball bearings.

4. A programmer must use a _____ _____ to map out specific
   locations on a workpiece.

5. Polar coordinates require that positions be identified by defining both a(n) _____ and
   a(n) _____ from the origin to a specific location.

6. Comments must be contained within _____ so the MCU knows not to read them when
   executing a program.

7. In CNC programming, the _____ character is called an end-of-block character.

Short Answer

8. List three characteristics or advantages of using linear guided ways.

_____

_____

_____

9. List four reasons why Acme lead screws are not used in CNC machines.

_____

_____

_____

_____

10. What are the two methods used in programming to define a workpiece position?

_____

_____

11. Explain a G-code.

_____

_____

12. Explain an M-code.

_____

_____

True or False

13. When a machine has an ATC, it can be classified as a "center."   _____

14. V-type and dovetail ways are most commonly found on CNC machines.   _____

15. A rectangular coordinate system is the same as a Cartesian coordinate system.   _____

16. A polar coordinate system uses the X- and Y-axes to define locations.   _____

17. Incremental positioning is cumulative from one set location.   _____

18. Modal codes are active until the machine is told to do otherwise.   _____

Identification Matching and Sketching

19. Make a sketch of the Cartesian coordinate system. Label the X- and Y-axes, each quadrant, and the origin.

20. Identify the X and Y coordinates on the grid you sketched using absolute programming.

21. Identify the X and Y coordinates on the grid you sketched using incremental programming. The tool initially is positioned at the origin.

Multiple Choice

22. The main difference between a CNC milling machine and a CNC milling center is the
   a. power knee.
   b. power quill.
   c. automatic tool changer.
   d. MCU.
   e. programmable Z-axis.

23. The best type of drive screw for use on a CNC machine is the
   a. acme screw.
   b. buttress screw.
   c. square screw.
   d. ball screw.
   e. round screw.

24. Servo motors are great for providing power to the axis because of their ability to
   a. support and move heavy objects.
   b. track how far the axis has moved.
   c. use less power.
   d. run for long periods of time without maintenance.
   e. None of the above

25. Movements in the "north" direction are given positive values and movements in the "south" direction are given negative values. What axis is being referenced?
   a. X
   b. Y
   c. Z
   d. A
   e. C

26. In incremental positioning, the last point of reference is called the
   a. absolute point.
   b. end of block.
   c. beginning point.
   d. point of origin.
   e. Both c. and d.
   f. None of the above

SECTION 8: Computer Numerical Control

UNIT 2	Introduction to CNC Turning

Name: _____   Date: _____

Score: _____   Text pages 578–592

Learning Objectives

After completing this unit, you should be able to:

- Identify and describe CNC turning machine types
- Identify parts of CNC turning machines
- Describe the machine axes used for turning
- Identify and describe toolholding and tool-mounting devices and their application for CNC turning
- Identify and describe workholding devices and their application for CNC turning

Carefully read the unit, and then answer the following questions in the space provided.

Fill in the Blank

1. Swiss turning centers slide the stock in and out of the spindle for _____ -axis motion.
2. Live tools are _____ _____ that enable a turning center to perform milling and off-center hole-work.
3. A _____ _____ threads onto the external threads on the collet chuck's end and constricts the collet when tightened by forcing it into the tapered bore.
4. _____ clamps are usually used to mount tools in gang-tool-type machines.
5. CNC _____ _____ consist of a straight round shank with a tapered bore.

Short Answer

6. List three advantages of a gang tool holder machine configuration.

   _____
   _____
   _____

7. List three factors to consider when selecting a workholding device.

   _____
   _____
   _____

8. Describe the manufacturing cell layout concept.

   _____
   _____

9. Why would a machinist use a subspindle?

_____

_____

10. Describe two machining operations that can be performed by live tooling.

_____

_____

11. Why do drill bushings have a tolerance of $\pm.001"$?

_____

_____

12. List three types of bar pullers.

_____

_____

_____

True or False

13. Turning machines do not have an X-axis.   _____

14. The spindle bore diameter limits the size of material used in collets.   _____

15. Mounting and dismounting is performed by rotating a single screw on a VDI-type toolholder.   _____

16. A CNC lathe has no ATC; a tool post is mounted on a cross slide, much like
    that of a manual lathe.   _____

17. Swiss turning machines use guide collets to stabilize and prevent the part
    from deflecting and vibrating.   _____

18. The spindle RPM is programmed the same for all toolholding devices.   _____

Multiple Choice

19. When creating turned parts, there is always the problem of not being able to machine the end of the workpiece
    held in the workholding device. What is the solution to this problem?
    a. Remove workpiece and remount
    b. Machine workpiece before mounting
    c. Use a subspindle
    d. Use a collet system with a through-bore spindle
    e. Use live tooling

20. What type of machine typically uses a flat bed design and is fitted with a carriage where tools are installed?
    a. Turret-type machine
    b. Gang tool machine
    c. Carousel tool machine
    d. Quick-change tool machine
    e. None of the above

21. If a machinist was machining thin-walled, hollow parts or tubing, what would be the best workholding device?

    a. Collet

    b. Three-jaw chuck

    c. Four-jaw chuck

    d. Faceplate

    e. Power chuck

22. To determine whether or not left-hand holders or right-hand holders are required for a turning application, what factor(s) need to be considered?

    a. Can the insert fit into the shape of the contour without clearance issues?

    b. Which direction will the tool be cutting, toward the headstock or away from the headstock?

    c. Will the tool be mounted in the machine upside down or right side up?

    d. The clearance of the toolholder and workholding device

    e. All of the above

UNIT 3 | CNC Turning: Programming

Name: _____ Date: _____

Score: _____ Text pages 593–619

Learning Objectives

After completing this unit, you should be able to:

- Identify basic G- and M-codes used for CNC turning
- Define and explain linear interpolation for CNC turning
- Define and explain circular interpolation for CNC turning
- Describe radial and diametral programming
- Describe facing operations for CNC turning
- Describe CNC rough turning operations
- Describe CNC finish turning operations
- Describe threading operations for CNC turning machines
- Describe tapping operations for CNC turning machines
- Describe various canned cycles for CNC turning applications
- Define and explain the principles of tool nose radius compensation (TNRC) for CNC turning

Carefully read the unit, and then answer the following questions in the space provided.

Fill in the Blank

1. A straight path of CNC motion to be taken between two programmed points can be used for straight turning or for taper turning. This is called _____ _____.

2. An arc of CNC cutting-tool motion to be created between two points in either the clockwise or counterclockwise direction is called _____ _____.

3. For quick positioning, a motion called _____ _____ is used.

4. To simplify repetitive and tedious roughing, finishing, and threading operations, a _____ cycle can be used.

5. A _____ must be programmed to compensate for a tool's radius on the left of the part while in motion and a _____ when on the right.

6. For circular interpolation, the characters _____ and _____ are used to define the arc center point in the X- and Z-axes.

Short Answer

7. List the three main motion types used in CNC programming.

_____

_____

_____

8. What is the TNRC and explain its use in CNC programming.

_____

_____

9. What is the common canned drill cycle code for CNC turning machines, and what letters are associated with the programming of the cycle?

_____

_____

10. What steps are taken at the beginning of a program to make a safe start and set default codes?

_____

_____

11. Why do most programmers prefer to space sequence numbers apart numerically between line numbers?

_____

_____

True or False

12. Code G00 is used for linear movement.   _____

13. There is a trade-off between too deep a cut and tool wear.   _____

14. The face of the workpiece is usually the Z-zero of the part.   _____

15. Facing cuts are typically performed from the inside of the workpiece toward the outer surface of the workpiece.   _____

16. There is only one way to program an arc on a CNC lathe.   _____

17. There are two ways to identify the feed rate on a CNC lathe.   _____

Computations

18. What is the pitch of 6–32 thread?

_____

_____

19. What is the feed rate for a 1/2-13 rigid tap operation?

_____

_____

20. To perform a facing cut using a tool with a nose radius of 1/64, what is the programmed X-axis end point at the center of the part?

_____

_____

21. Explain each function of the following coding: G90 G20 M4 S1200 T0202.

_____

_____

_____

_____

_____

Multiple Choice

22. What code would be used for rapid, non-cutting axis movements?

    a. G01

    b. G19

    c. M0

    d. G00

    e. M05

23. When using cutter compensation, what code moves the tool to the right of the toolpath?

    a. G42

    b. G43

    c. G40

    d. G45

    e. None of the above

24. What code is used for finishing canned cycles?

    a. G75

    b. G73

    c. G72

    d. G70

    e. G71

UNIT 4 | CNC Turning: Setup and Operation

Name: _____   Date: _____

Score: _____   Text pages 620–632

Learning Objectives

After completing this unit, you should be able to:

- Describe CNC machine modes
- Describe the work coordinate system (WCS) for CNC turning
- Describe the machine coordinate system (MCS) for CNC turning
- Describe the homing procedure and purpose
- Describe workpiece offsets for CNC turning
- Describe tool geometry offsets for CNC turning
- Describe tool wear offsets for CNC turning
- Describe tool nose radius (or diameter) offsets
- Describe tool quadrant settings for TNRC
- Define the three basic methods for loading programs into the MCU
- Describe program prove-out procedures

Carefully read the unit, and then answer the following questions in the space provided.

Fill in the Blank

1. The machine control mode used for entering and storing programs is called the _____ mode.
2. Pressing and holding the _____ _____ button causes constant axis motion.
3. The _____ _____ button begins the active CNC program and is colored green on many machines.
4. Typically, _____ of minimum clearance should be maintained between the cutting tool and the workholding device when positioning.
5. The _____ defines the workpiece coordinates and the _____ defines the machine coordinates.

Short Answer

6. What are the five modes in which a CNC machine can be operated?

_____

_____

_____

_____

_____

7. Explain the term *workshift*.

_____

_____

8. What can be used to adjust for tool wear during a production run?

_____

_____

9. Explain the process of inspection of cutting-tool squareness with a dial indicator.

_____

_____

10. Explain three ways to enter a program in the MCU.

_____

_____

_____

11. What is the best way to prove-out a program?

_____

_____

True or False

12. Some buttons on the control panel do not have labels.   _____

13. Programs entered using MDI are stored in memory.   _____

14. There are override controls for speeds and feeds.   _____

15. The CNC drawbar determines the collet holding pressure.   _____

16. The first step when operating a CNC is the homing process.   _____

17. The MCS always changes after each workpiece is mounted.   _____

Computations

18. Using the method described in this unit, if a tool is touched off on a 1.250" diameter with a .015" feeler gage, what is the value that will be entered for measurement into the geometry offset page?

_____

_____

19. Using the method described in this unit, if a tool is touched off on the Z-zero surface of the part with a .015" feeler gage, what is the value that will be entered for measurement into the geometry offset page?

_____

_____

Multiple Choice

20. A measurement reveals that each diameter on the workpiece measures .0008" larger than the desired size. The current X-axis geometry offset for the tool is 4.6789". Does the machinist
    a. add the .0008" to the offset?
    b. subtract the .0008" off the offset?
    c. double the .0008" and add to the offset?
    d. change the WCS location?
    e. add .0004" to the offset?

21. In certain cases, instead of actually storing a program in the MCU memory, the program is fed from a PC to the control line by line as the machine runs the program. This is called
    a. drip feeding.
    b. manual data input.
    c. direct numerical control.
    d. USB download.
    e. Both a. and c.

22. When mounting cutting tools in toolholders, the machinist must check for
    a. clearance between cutting tool and holder.
    b. amount of cutting tool exposed.
    c. squareness of the cutting tool to the hold axis.
    d. tightness of cutting tool in holder.
    e. All of the above
    f. None of the above

23. The process of homing a CNC machines is to
    a. re-zero all axes.
    b. establish an MCS.
    c. recall the position of the WCS.
    d. establish travel limits.
    e. All of the above
    f. None of the above

SECTION 8: Computer Numerical Control

UNIT 5 | Introduction to CNC Milling

Name: _____ Date: _____

Score: _____ Text pages 633–647

Learning Objectives

After completing this unit, you should be able to:

- Identify and describe CNC milling machine types
- Describe the machine axes used for milling
- Describe the two major types of ATCs
- Identify and describe workholding devices for CNC milling
- Identify and describe toolholding devices used for CNC milling

Carefully read the unit, and then answer the following questions in the space provided.

Fill in the Blank

1. A _____ _____ is gripped by a drawbar mechanism in the machine spindle to secure toolholders.

2. Machining centers are separated into two major classes: _____ _____ and _____ _____.

3. The use of _____ in machine design greatly minimizes sliding surface wear, reduces friction, and allows for very high accuracies due to a zero-clearance preloaded ball bearing design.

4. Machining center spindles use an NMTB taper size and have a taper of _____ per foot.

5. The combination of all machining operations required to fabricate a part is called a _____ _____.

Short Answer

6. What are two advantages of using a shrink-fit toolholder?

   _____

   _____

7. Name the two most commonly used toolholder flange types.

   _____

   _____

8. List five machining operations that a milling-type CNC can perform.

   _____

   _____

   _____

   _____

   _____

9. Explain the procedure for securing an endmill in an end holder.

_____

_____

10. Describe how to reduce the amount of time needed to align a part to the machine axis.

_____

_____

True or False

11. CNC machining centers are defined by their knee size. _____

12. The swing-arm tool changer is much faster then the carousel changer. _____

13. A difference between a CAT and BT toolholder is the internal thread series. _____

14. Some CNC milling machines do not use either CAT or BT flange holders. _____

15. A process chart is used by programmers and machine operators. _____

Identification Matching

16. Select the best workholding device for the given machining operations.

   a. Clamps   _____   **1.** For holding thin, flexible parts flat

   b. Vises   _____   **2.** Hold and locate cylindrical parts

   c. Tombstones   _____   **3.** Holding extremely large or oddly shaped workpieces

   d. Vacuum plate   _____   **4.** Device specifically designed to accommodate a specific part

   e. Custom fixture   _____   **5.** Square part that can't bow

   f. Index fixture   _____   **6.** Uses two or more workholding tooling plates

   g. Pallet system   _____   **7.** Multiple working surfaces or workpiece mounted at one time

Multiple Choice

17. On a CNC machining center, tools are held in the spindle by the
   a. taper.
   b. toolholder.
   c. retention knob.
   d. arbor.
   e. flange.

18. A CAT-40 holder uses the CAT flange and a size
   a. 50 taper.
   b. 40 taper.
   c. 30 taper.
   d. 45 taper.
   e. 35 taper.

19. The best general toolholder for endmills to reduce run-out is a
   a. solid toolholder with set screw.
   b. shrink-fit holder.
   c. collet chuck.
   d. arbor chuck.
   e. drill chuck holder.

20. The most commonly used workholding devices for milling are
    a. toe clamps.
    b. parallel jaw vises.
    c. indexing fixtures.
    d. pallet plates.
    e. tombstones.

UNIT 6 | CNC Milling: Programming

Name: _____   Date: _____

Score: _____   Text pages 648–682

Learning Objectives

After completing this unit, you should be able to:

- Identify basic G- and M-codes used for CNC milling
- Define and explain linear interpolation for CNC milling
- Define and explain circular interpolation for CNC milling
- Explain the arc center method for circular interpolation
- Explain the radius method for circular interpolation
- Describe facing operations for CNC milling
- Describe types of two-dimensional CNC milling
- Define and explain drilling and tapping canned cycles for milling
- Define and explain cutter radius compensation (cutter comp) for milling

Carefully read the unit, and then answer the following questions in the space provided.

Fill in the Blank

1. A _____ code causes a full stop and a _____ code causes an optional stop.
2. The move used to activate and cancel cutter comp must be in a _____ move.
3. The _____ code sets the XY plane during the safe-start portion of a program.
4. The letters _____ and _____ are used to generate arcs and circles.
5. The _____ surface is the point that all other surfaces will be referenced from.

Short Answer

6. What command would be given to turn on the spindle in a counterclockwise direction at 250 RPM?

_____

_____

7. What command would be given to execute a tool change to tool #8?

_____

_____

8. Write a block of code to machine a clockwise .2" radius with a start point of X1.0, Y1.0 and an endpoint of X1.2, Y1.2 using the radius method.

_____

_____

9. What command would be used to activate tool height offset compensation for tool #5?

_____

_____

10. Explain the differences between single-pass, fast-peck, and deep-hole drilling canned cycles.

_____

_____

11. Explain each function of the coding: G0 G54 X5.5 Y3.0.

_____

_____

_____

_____

True or False

12. There is a standard format for CNC programming.   _____

13. The values of the H-word and the T-word must always match or the machine will provide an
    alarm message.   _____

14. Rapid traverse out of a hole will cause dulling of drills.   _____

15. G74 is used for right-hand tapping in canned cycles.   _____

16. A D-word must also be programmed to activate a tool radius offset value that is stored in the MCU. _____

17. Canned cycles can be used to simplify machining operations.   _____

Identification Matching

18. Match the proper code to the given command.

a. G95	_____	1. Spindle off
b. G90	_____	2. Coolant on
c. G20	_____	3. Spindle on clockwise
d. M05	_____	4. Inches per revolution
e. M08	_____	5. Absolute positioning
f. M03	_____	6. Sets to inches

Multiple Choice

19. The code for feed rates expressed in inches per minute (IPM) is

    a. G17.
    b. G90.
    c. G94.
    d. G95.
    e. G93.

20. At the beginning of each block of code each block is labeled with a
 a. semicolon.
 b. colon.
 c. parentheses.
 d. sequence number.
 e. sequence space.

21. In CNC mill programming, it is standard practice to establish a safety zone above the workpiece or workholding device called a clearance plane. Typically, a clearance plane of _____ is used.
 a. .625"
 b. 1.00"
 c. .100"
 d. .200"
 e. All the above

22. The code used to stop all programmed functions during a machining operation is a
 a. M01.
 b. M00.
 c. M03.
 d. M05.
 e. M09.

23. What letter is associated with the tool height offset geometry?
 a. I
 b. J
 c. K
 d. H
 e. T

UNIT 7	CNC Milling: Setup and Operation

Name: _____ Date: _____

Score: _____ Text pages 683–697

Learning Objectives

After completing this unit, you should be able to:

- Describe CNC machine modes for CNC milling
- Describe the work coordinate system (WCS) for CNC milling
- Describe the machine coordinate system (MCS) for CNC milling
- Describe the homing procedure and purpose for CNC milling
- Describe workpiece offsets for CNC milling
- Describe tool geometry offsets for CNC milling
- Describe tool wear offsets for CNC milling
- Describe cutter radius compensation offsets
- Define the three basic methods for loading programs into the MCU
- Describe program prove-out procedures for CNC milling

Carefully read the unit, and then answer the following questions in the space provided.

Fill in the Blank

1. A(n) _____ control is used to slow rapid motion while a program is running or to vary the jogging rate.
2. When workpiece clamps are used, the machinist must plan for clamp _____.
3. The _____ _____ establishes the workpiece location.
4. When using a tombstone mounting system, numerous work coordinate systems can be programmed using the _____ series of codes.
5. The _____ axis work offset is usually the first to be established.
6. To locate a workpiece that has a hole in the center, a _____ _____ would be best to use.
7. The machine may be run at full programmed feed, speed, and at 100% of its rapid capabilities when in the _____ mode.

Short Answer

8. Using the method described in this unit, if a tool is touched off the Z-zero surface with a .015 feeler gage and the absolute (work) position register says the position is 4.5432, what is the total length of the tool?

_____

_____

9. Using the method described in this unit, if the spindle face is touched off of the Z-zero surface of the part with a 2.0000 gage block, and the machine position register says the position is −18.5678, what is the total work offset for the Z-axis? (Include proper sign.)

_____

_____

10. Assume that a tool radius value of .2500" is entered initially in the tool offset page and that 200 parts have been made within tolerance. Then the tool begins producing parts that are too large in their overall length and width dimensions by a total of .0015". What should be done to correct the problem?

_____

_____

11. What is the most common way to load a program into memory?

_____

_____

True or False

12. The homing process locates the WCS of the machine. _____

13. The length of a tool is measured from the table line on the toolholder's taper to the tool tip. _____

14. The entered radius value of the tool is ignored unless cutter comp is activated. _____

15. Adjustments can be made to either the tool height or the tool radius offset. _____

16. Graphic simulation is the least best way to test a program for errors. _____

Multiple Choice

17. If a machinist wants to manually move the X-axis, what mode should be selected?
    a. MDI
    b. Auto
    c. Jog
    d. Edit
    e. Zero-reference

18. If a machinist wants to save a program into the MCU, what mode should be selected?
    a. MDI
    b. Auto
    c. Jog
    d. Edit
    e. Zero-reference

19. In an emergency or collision with a(n) object, what button should be pushed?
    a. Feed hold
    b. Feed override
    c. Rapid override
    d. Red emergency stop button
    e. All of the above

20. No matter which device is used to secure a workpiece when multiple parts are to be machined, it is important to
   a. use the same torque on clamp fasteners.
   b. determine the proper Z-axis zero.
   c. check for workholding device clearance.
   d. use the same torque on vise screws.
   e. All of the above

SECTION 8: Computer Numerical Control

UNIT 8	Computer-Aided Design and Computer-Aided Manufacturing

Name: _____ Date: _____

Score: _____ Text pages 698–706

Learning Objectives

After completing this unit, you should be able to:

- Describe the basic applications of CAD
- Describe the basic applications of CAM
- Explain the uses of and be able to recognize wireframe drawings
- Explain the uses of and be able to recognize solid model drawings
- Explain the uses of and be able to recognize surface drawings
- Describe the basic principles of toolpath creation
- Describe basic toolpath types
- Describe the basic principles of post-processing

Carefully read the unit, and then answer the following questions in the space provided.

Fill in the Blank

1. The final step called _____ is to take all of the defined toolpath data and allow the CAM software to write a CNC program.

2. _____ _____ surface milling is where CAM is most useful.

3. In order to create toolpaths, _____ are selected in the order of machining operation.

4. A group of selected entities is called a(n) _____.

5. The ridges left from a ball mill are called _____.

Short Answer

6. Explain the process of milling a pocket in a block of aluminum.

_____

_____

7. List the three types of CAD drawing forms.

_____

_____

_____

8. List three reasons to simulate a post-processed program.

_____

_____

_____

9. List four toolpath examples for each milling and turning operation.

_____

_____

_____

_____

True or False

10. Turning operations are often programmed in longhand without using CAM software.   _____

11. CAD/CAM eliminated the use of paper to convey the drawing.   _____

12. CAD software uses post-processing operations.   _____

13. Post-processors write code generically for all machine control formats.   _____

14. For surface milling, boundaries are set for tool travel.   _____

15. Not all CAM software packages require the part to be drawn in a separate CAD software system.   _____

Multiple Choice

16. Which endmill is best for 3-D contour milling?
     a. Center-cutting endmill
     b. Four-flute endmill
     c. Two-flute endmill
     d. Two-flute ball mill
     e. Shell endmill

17. What part of the CNC programming process involves creation of a CNC program?
     a. Toolpath creation
     b. Geometry creation
     c. Post-processing
     d. Wire framing
     e. Solid modeling

18. To improve the surface roughness remaining from the cusp height, the user may also desire to switch to a different
     a. pattern direction of the tool during finishing.
     b. change of cutting tools.
     c. change of speed and feeds.
     d. adjustment of tool offsets.
     e. adjustment of tool height offsets.

Drill Drift

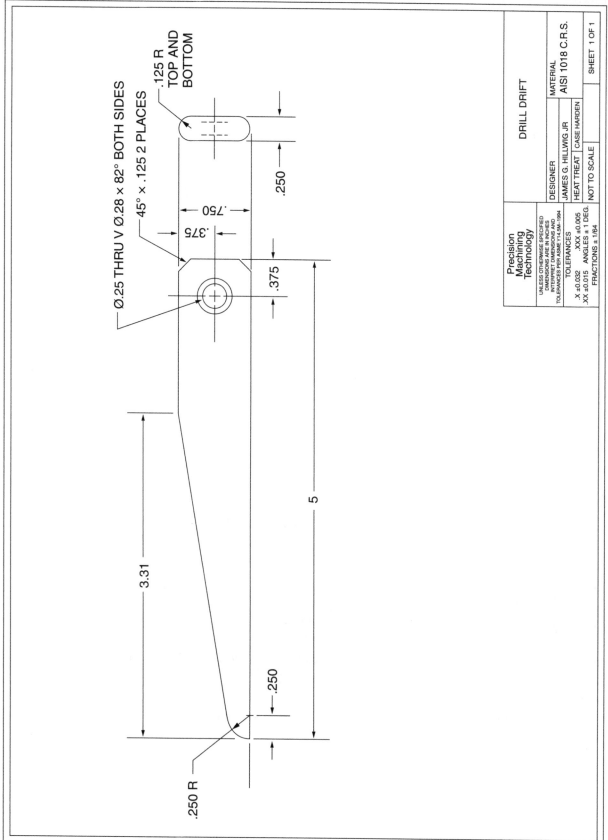

.125 R
TOP AND
BOTTOM

Ø.25 THRU V Ø.28 × 82° BOTH SIDES

45° × .125 2 PLACES

.250

.750

.375

.375

5

3.31

.250

.250 R

Precision Machining Technology	DRILL DRIFT		
UNLESS OTHERWISE SPECIFIED DIMENSIONS ARE IN INCHES INTERPRET DIMENSIONS AND TOLERANCES PER ASME Y14.5M—1994	DESIGNER	MATERIAL	
	JAMES G. HILLWIG JR	AISI 1018 C.R.S.	
TOLERANCES .X ±0.032 .XXX ±0.005 .XX ±0.015 ANGLES ± 1 DEG. FRACTIONS ± 1/64	HEAT TREAT CASE HARDEN		
	NOT TO SCALE	SHEET 1 OF 1	

159

NIMS Skills Practiced

- Job Process Planning
- Layout
- Benchwork
- Drill Press

Equipment

Power saw, bench vise, drill press

Tools Needed

3" to 6" dividers	Scribe	6" steel rule
Layout dye	Prick punch	Center punch
Hammer	Hacksaw	File
File card	Abrasive cloths	1/4" drill bit
82° countersink		

Material Needed

- 1/4" × 3/4" × 5-1/8" mild steel

Safety

1. Follow all shop safety rules given to you by your instructor and be sure to pass any safety tests given before using shop equipment.
2. Follow all safety rules for power equipment.
3. Be careful of sharp tools when using them.
4. Make sure the file has a handle on it.

Order of Operations

1. Look over blueprint to determine tolerances, tools, and material needed.
2. Gather tools needed, figure out RPMs for required tools, and write down their RPMs on sheet.
3. Measure, mark, and cut stock 1/8" longer than length on print so you can square up one end.
4. Square one end and deburr for layout.
5. Paint with dye and lay out part.
6. Drill 1/4"-diameter hole, then countersink both sides with 82° countersink.
7. Cut with hacksaw close to your layout lines, staying 1/32" away.
8. Finish by filing the outside of part to layout lines, being careful to follow print dimensions.
9. File radii where indicated on print.
10. Polish with abrasive cloths.
11. Inspect the part to print dimensions; record on inspection sheet and turn in.

How to Determine the RPMs

Tool Being Used	Surface Feet per Minute	Formula	RPM
1/4" drill bit	80	$\dfrac{4 \times 80}{.250}$	
82° countersink	80	$\dfrac{4 \times 80}{\text{Diameter of Countersink}}$	

Student Name: _____ Date Submitted: _____

Class: _____ Total Hours on Job: _____

	Print Dimensions AND Tolerances	Student's Inspection	Instructor's Inspection	Instructor Comments
1	Length 5" Long ± 1/64"			
2	Width 1/4" Wide ± 1/64"			
3	Height 3/4" High ± 1/64"			
4	Taper 3.31" Long ± 1/64"			
5	Front Radius ± 1/64"			
6	Height to Front Radius ± 1/64"			
7	Distance from End to Radius ± 1/64"			
8	Rear Chamfer .125" × 45° (2 places)			
9	Hole 1/4" Diameter ± 1/64"			
10	82° Countersink Ø.280" Deep ± 1/64"			
11	Centered .375" × .375" 1/4" Hole			
12	Top Radius ± 1/64"			
13	Bottom Radius ± 1/64"			
14	Is the part free of burrs and does it meet finish requirements?			
15	Is the part properly identified?			

PROJECT GRADE: _____

"T" Slot Mill Cleaner

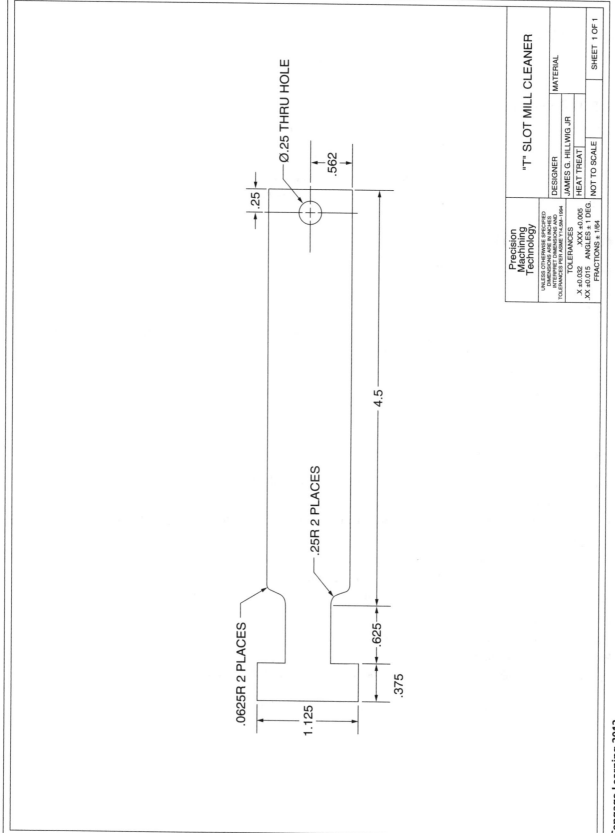

Ø.25 THRU HOLE

.562

.25

4.5

.25R 2 PLACES

.0625R 2 PLACES

1.125

.625

.375

Precision Machining Technology	"T" SLOT MILL CLEANER	
	DESIGNER	MATERIAL
UNLESS OTHERWISE SPECIFIED DIMENSIONS ARE IN INCHES INTERPRET DIMENSIONS AND TOLERANCES PER ASME Y14.5M–1994	JAMES G. HILLWIG JR	
TOLERANCES	HEAT TREAT	
X ±0.032 XXX ±0.005		
XX ±0.015 ANGLES ± 1 DEG.	NOT TO SCALE	SHEET 1 OF 1
FRACTIONS ± 1/64		

NIMS Skills Practiced

- Job Process Planning
- Layout
- Benchwork

Equipment

Power saw, bench vise, drill press

Tools Needed

3" to 6" dividers   Scribe   12" combination set
Layout dye   Prick punch   Center punch
Hammer   Hacksaw   File
File card   Abrasive cloths   Radius gages
Center drill   1/4" drill   Countersink

Material Needed

- 1/8" × 1-3/16" × 4-1/2" aluminum or CRS

Safety

1. Follow all shop safety rules given to you by your instructor and be sure to pass any safety tests given before using shop equipment.
2. Follow all safety rules for power equipment.
3. Be careful of sharp tools when using them.
4. Make sure the file has a handle on it.

Order of Operations

1. Look over blueprint to determine tolerances, tools, and materials needed.
2. Gather tools needed.
3. Measure, mark, and cut stock 1/8" longer than length on print so you can square up one end.
4. Using file and square, square up one end and deburr for layout.
5. Paint with dye and lay out part.
6. Cut with hacksaw close to your layout lines, staying 1/32" away.
7. Finish by filing the outside of part to layout lines, being careful to follow print dimensions.
8. File radii where indicated on print.
9. Center drill, drill, and countersink 1/4" hole.
10. Polish with abrasive cloths.
11. Inspect the part to print dimensions; record on inspection sheet and turn in.

Student Name: _____ Date Submitted: _____

Class: _____ Total Hours on Job: _____

	Print Dimensions AND Tolerances	Student's Inspection	Instructor's Inspection	Instructor Comments
1	Is the part free of burrs and does it meet finish requirements?			
2	Is the part properly identified?			
3	6" ± 1/64"			
4	1/8" ± 1/64"			
5	1" ± 1/64"			
6	1/4" (1 place)			
7	7/16" (2 places)			
8	.500"			
9	1/16" R ± 1/64" (4 places)			
10	1/4" R ± 1/64" (2 places)			
11	1/2" R ± 1/64"			
12	1" ± 1/64" (1/2" radius end)			
13				
14				
15				
16				
17				
18				
19				
20				

PROJECT GRADE: _____

Soft Jaws for Bench Vise

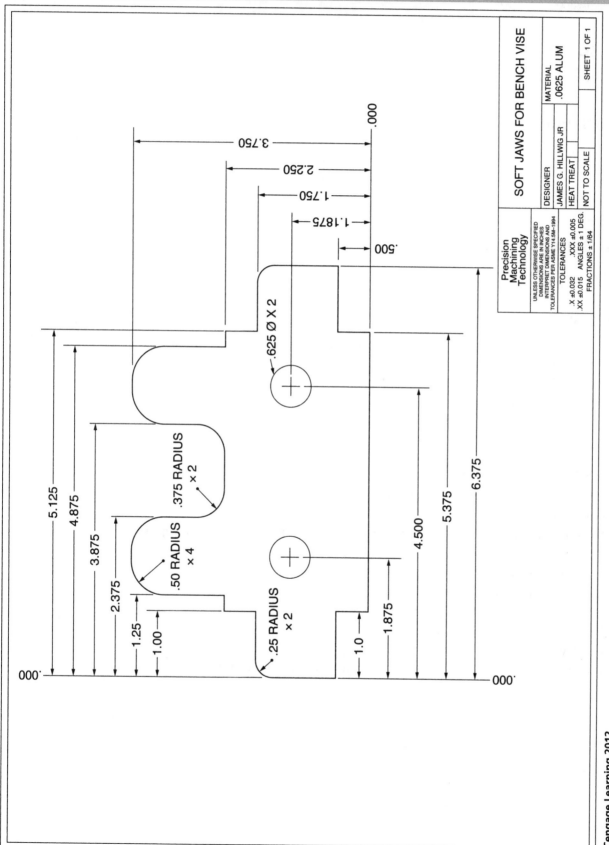

SOFT JAWS FOR BENCH VISE

DESIGNER — JAMES G. HILLWIG JR

MATERIAL — .0625 ALUM

HEAT TREAT

NOT TO SCALE — SHEET 1 OF 1

Precision Machining Technology

UNLESS OTHERWISE SPECIFIED
DIMENSIONS ARE IN INCHES
INTERPRET DIMENSIONS AND
TOLERANCES PER ASME Y14.5M-1994

TOLERANCES
.X ±0.032   .XXX ±0.005
.XX ±0.015   ANGLES ± 1 DEG.
FRACTIONS ± 1/64

.625 Ø X 2

.375 RADIUS × 2

.50 RADIUS × 4

.25 RADIUS × 2

3.750
2.250
1.750
1.1875
.500

5.125
4.875
3.875
2.375
1.25
1.00

6.375
5.375
4.500
1.875
1.0

.000
.000

NIMS Skills Practiced

- Job Process Planning
- Layout
- Benchwork

Equipment

Power saw, bench vise

Tools Needed

3" to 6" dividers	Scribe	12" combination set
Layout dye	Prick punch	Center punch
Hammer	Hacksaw	File
File card	Abrasive cloths	Radius gages

Material Needed

- 1/16" × 4" × 6-1/2" aluminum

Safety

1. Follow all shop safety rules given to you by your instructor and be sure to pass any safety tests given before using shop equipment.
2. Follow all safety rules for power equipment.
3. Be careful of sharp tools when using them.
4. Make sure the file has a handle on it.

Order of Operations

1. Look over blueprint to determine tolerances, tools, and material needed.
2. Gather tools needed.
3. Measure, mark, and cut stock 1/8" longer than length on print so you can square up one end.
4. Using file and square, square up one end and deburr for layout.
5. Paint with dye and lay out part.
6. Cut with hacksaw close to your layout lines, staying 1/32" away.
7. Finish by filing the outside of part to layout lines, being careful to follow print dimensions.
8. File radii where indicated on print.
9. Polish with abrasive cloths.
10. Inspect the part to print dimensions; record on inspection sheet and turn in.

Student Name: _____ Date Submitted: _____

Class: _____ Total Hours on Job: _____

	Print Dimensions AND Tolerances	Student's Inspection	Instructor's Inspection	Instructor Comments
1	Is the part free of burrs and does it meet finish requirements?			
2	Is the part properly identified?			
3	**Length** 6.375"			
4	**Width** 3-3/4"			
5	**Holes** 5/8" Diameter			
6	2-5/8" Apart			
7	**Bottom** 4.375" Length			
8	1" Length (2 places)			
9	.50" Height (2 places)			
10	**Sides** 1.25" Height			
11	.25" R (2 places)			
12	.50" Height (2 places)			
13	1" Length (2 places)			
14	1.500" Height (4 places)			
15	.250" Length (2 places)			
16	.50" Height (2 places)			
17	**Top** .5" R (4 places)			
18	1-1/8" (2 places)			
19	3/8" R (2 places)			
20	1-1/2" Length			

PROJECT GRADE: _____

C-Clamp

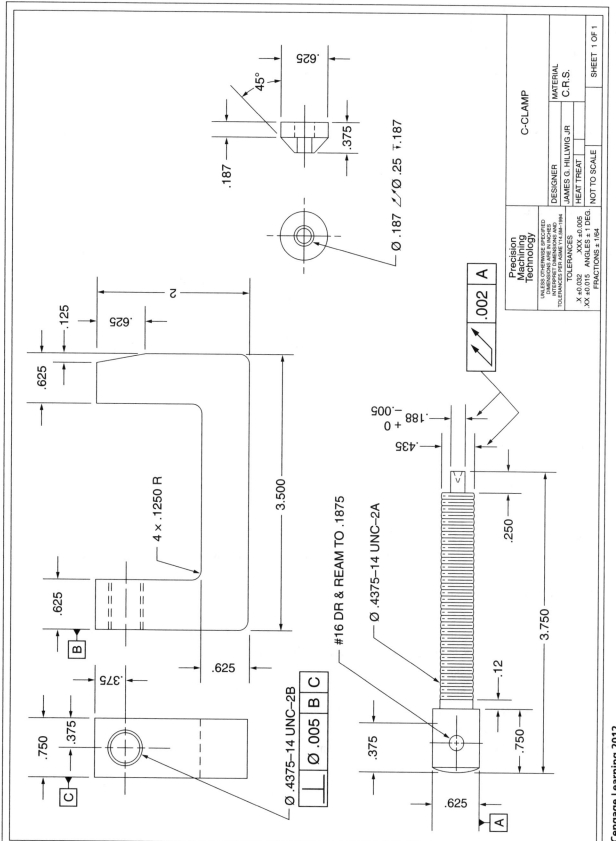

Ø .187 ⟋ Ø .25 ⊤ .187

45°

.625

.187

.375

2

.125

.625

.625

3.500

4 × .1250 R

.625

.625

.375

.750

.375

B

C

| ⟂ | Ø .005 | B | C |

Ø .4375–14 UNC–2B

#16 DR & REAM TO .1875

Ø .4375–14 UNC–2A

.188 +0 −.005

.435

| ∠ | .002 | A |

.250

3.750

.12

.375

.750

.625

A

Precision Machining Technology

C-CLAMP

DESIGNER
JAMES G. HILLWIG JR

MATERIAL
C.R.S.

HEAT TREAT

NOT TO SCALE

SHEET 1 OF 1

UNLESS OTHERWISE SPECIFIED
DIMENSIONS ARE IN INCHES
INTERPRET DIMENSIONS AND
TOLERANCES PER ASME Y14.5M–1994

TOLERANCES
.X ±0.032   .XXX ±0.005
.XX ±0.015   ANGLES ± 1 DEG.
FRACTIONS ± 1/64

NIMS Skills Practiced

- Job Process Planning
- Layout
- Turning Operations: Chucking
- Drilling and Tapping
- Surface Grinding and Safety
- Vertical Milling and Safety
- Squaring up a Block

Equipment

Power saw, milling machine, lathe, surface grinder

Tools Needed

Scribe	6" steel rule	Layout dye
Dead blow hammer	File	File card
Various drill bits	7/16-14 tap	Stamps
90° countersink	Various sizes of endmills	Parallels
Edge finder	Center drill	Vise stop
Machinist square	Grinding wheel	Grinder vise
Tramming items for mill	Adjustable mill angle	Thread triangles
Various types of turning, threading, and facing tools	3/16" reamer	Magnetic sine plate

Material Needed

- 3/4" thick × 2" wide × 3-5/8" long CRS (for "C" body)
- 5/8" diameter × 4" long CRS (for screw)
- 5/8" diameter bar about 12" long (to safely hold cap while machining)
- 3/16" diameter × 1-1/2" long CRS (for screw handle)

Safety

1. Follow all shop safety rules given to you by your instructor and be sure to pass any safety tests given before using shop equipment.
2. Follow all safety rules for power equipment.
3. Be careful of sharp tools when using them.
4. Make sure the file has a handle on it.
5. Be sure all cutting tools are sharp and care is taken while handling them.
6. Be sure ways are locked on mill when necessary.
7. Be sure chuck is tightened and chuck wrench is never left in chuck.
8. Make sure stock is supported with center when necessary.
9. Ring grinding wheels before using them.
10. Make sure magnetic chuck is on when surface grinding.

Order of Operations

MILL

1. Look over blueprint to determine tolerances, tools, and material needed.
2. Gather tools needed, figure out RPMs for required tools, and write down their RPMs on sheet.
3. Measure, mark, and cut stock 3/4" thick × 2" wide. Saw 3-5/8" long.
4. Make sure milling machine is in tram.
5. Deburr all over.
6. Place piece in mill vise on parallels with sawed end sticking out; tap down with dead blow hammer to seat part on parallels, then mill just enough off the end to clean it up.
7. Deburr the end you just milled, flip part 180° on parallels in vise, tap down with dead blow hammer to seat part on parallels, and mill to length.
8. Lay out part.
9. Place part in vise on parallels against vise stop, tap down with dead blow hammer to seat part on parallels, and set vise stop on left side of part.
10. Insert edge finder in collet, set RPMs to 1100, and edge find the X-axis and the Y-axis. Be sure to allow for 1/2 grind stock on the X-axis.
    a. X-axis should read −.1075 (this includes 1/2 of the edge finder and 1/2 of the grind stock).
    b. Y-axis should read .1000 (with the tolerance on the print, the 2" stock size should leave plenty of room for grinding).
11. Place center drill in chuck, and move in 5/8" + .015 + 1/8" on the X-axis.
12. Take 2"—(5/8" + .015 + 1/8"); this will be your Y-axis location.
13. Lower center drill to make sure it aligns with layout lines.
14. If location is correct, center drill hole at 1200 RPMs, change to 1/4" drill, set the RPMs, and drill through part.
15. Remove part and deburr, flip part end for end against your stop and repeat the center drill and drilling operations, then remove part and deburr. If done correctly, the holes should be perfect.
16. Now stand part up on parallels, holding the bottom of the "C" in the vise with holes you drilled just above the top of the vise jaws.
17. Place a 3/4" roughing endmill in a collet, keeping away from the 5/8" layout line on the side, and mill down until you reach the center of the hole.
18. Using a band saw, you can now cut to the center of the remaining hole from the top and slide the blade in the 3/4" slot you just cut along the bottom of the "C" to meet where you just cut. This will remove the bulk of the center.
19. Using a finishing endmill, place the part back in the vise and clean up the sides and bottom, leaving grind stock. DO NOT REMOVE THE RADIUS FROM THE CORNERS.
20. Now stand up on end against the vise stop, keeping the 2" side along vise jaws, and move to the 7/16-14 hole location.
21. Center drill, and drill with a letter "U" drill, countersink the hole, then power tap the hole with a 7/16-14 tap.
22. Trig out the angle; with the 1/8" and 5/8" locations of the angle laid out, set the opposite corner of the drilled end of the part on the mill angle, and mill the angle to layout lines. DO NOT remove lines; just touch them— this will leave grind stock. (Have instructor check setup before you begin your cut.)
23. Deburr part and hold with the bottom of the "C" up against the vise stop. Using a 3/8" endmill, cut a name pocket centered both ways .030 deep, 3/8" wide × 3/4" long.
24. Now hold in a bench vise and file the 1/8" radius on the bottom corners of the part.
25. Deburr, place on anvil, and using steel stamps, stamp your name in pocket.

* This part can be heat treated if your budget permits, but it is not necessary.*

SURFACE GRINDER

MAKE SURE MAGNETIC CHUCK IS ON WHEN SURFACE GRINDING!

26. Surface grind all sides square and parallel. Deburr after each side is ground.

27. Surface grind the face to clean it up, deburr, then flip the part to grind the opposite side.

28. Deburr all edges and place all the way down in grinder vise with ground sides against jaws, leaving one edge sticking out of side of vise so you can turn vise on its side.

29. Grind top of part to clean up and turn vise on side without removing or disturbing part while in vise and grind just enough to clean it up.

30. Remove from vise, deburr all edges, place ground side down on chuck with square blocking on each side, and grind bottom to clean it up.

31. Deburr and place last side up with square blocking on each side and clean up surface.

32. Now check sizes and hole location; grind them into location and part to size. Be sure to deburr and block part when needed.

33. Using vise and magnetic sine bar, calculate gage block height, set magnetic sine bar to height, place part in grinder vise, place on magnetic sine bar, TURN ON MAGNETS, and grind angle.

34. Using a sine bar and roll dimensions, check the angle. (Have your instructor help you.)

35. Dressing a radius on your grinding wheel: grind the inside radius on the clamp. (See instructor.)

36. Lay flat on grinder chuck and grind a .030 × 45° chamfer on all edges.

LATHE

37. Chuck 5/8" diameter × 4" long stock in lathe face until end is clean.

38. Mark faced end at .750; this will be the head of the screw.

39. File the convex on the face of the screw.

40. Turn part end for end, face to length, and center drill 3/16" diameter using a #3 center drill. The center-drilled hole remains in the part this time.

41. Turn .435 diameter back 3" to the .750 mark of the head.

42. Chase the thread on the screw.

43. Turn the .188 diameter back 1/4".

44. Set screw in mill or drill press and drill and ream the 3/16" hole in the head.

45. Chuck the 5/8" diameter bar for the cap in the lathe, face, center drill, and drill the 3/16" hole 1/2" deep.

46. Turn the 45° angle 3/16" back.

47. Using the cutoff tool, cut the cap off at 3/8".

48. Turn cap end for end, CAREFULLY place the cap in the chuck, and drill the 1/4" hole 3/16" deep.

49. Place the 3/16"-diameter stock in the lathe, face to length, and put a radius on each end.

50. Slide the 3/16"-diameter handle into screw head and smash ends in press or vise so it will not fall out.

51. Place screw into "C" and put cap on end of screw; using a socket-head cap screw, insert the end of the screw into the 1/4" hole of cap and the other end against the side of "C" and tighten the screw. This will mushroom out the center-drilled hole in the end of the screw to keep it on the screw.

52. Evaluate and turn in.

Student Name: _____ Date Submitted: _____

Class: _____ Total Hours on Job: _____

	Print Dimensions AND Tolerances	Student's Inspection	Instructor's Inspection	Instructor Comments
1	Is the part free of burrs and does it meet finish requirements?			
2	Is the part properly identified?			
3	3/4" ± .015 Thickness of "C"			
4	3 1/2" ± .015 Length of "C"			
5	2" ± .015 Height of "C"			
6	5/8" ± .015 Threaded Side of "C"			
7	5/8" ± .015 Bottom of "C"			
8	5/8" ± .015 Angle Side of "C"			
9	1/8" ± .015 Angle on "C"			You will NEED to use the comparator
10	5/8" ± .015 Angle on "C"			You will NEED to use the comparator
11	3/4" ± .015 Head of Screw Length			You will NEED to use the comparator
12	5/8" ± .015 Head of Screw			
13	.435 ± .002 Diameter of 7/16—14 Thread			
14	.12 ± .015 Diameter Behind Head, Unthreaded Portion of Screw			You will NEED to use the comparator
15	3/8" ± .015 Length of Cap			You will NEED to use the comparator
16	3/16" ± .015 Flat on Cap			You will NEED to use the comparator
17	5/8" ± .015 Diameter of Cap			
18	45° ± 1/2° Angle on Cap			You will NEED to use the comparator
19	1/8" ± .015 Radius Inside "C"			You will NEED to use the comparator
20	1/8" ± .015 Radius Outside "C"			You will NEED to use the comparator

PROJECT GRADE: _____

Lathe Dog

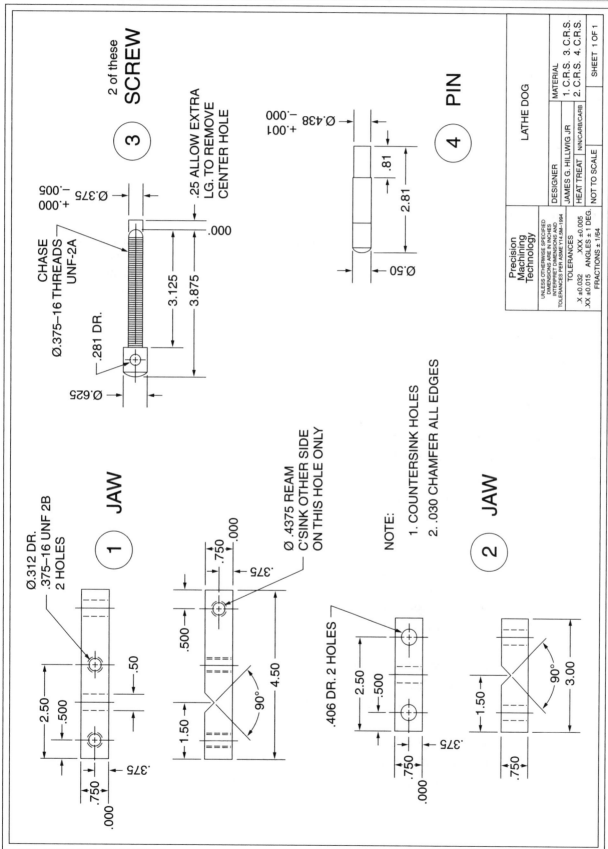

③ SCREW — 2 of these

Ø.375 +.000 -.005

CHASE
Ø.375–16 THREADS
UNF-2A

.281 DR.

Ø.625

3.125

3.875

.25 ALLOW EXTRA
LG. TO REMOVE
CENTER HOLE

.000

④ PIN

Ø.438 +.001 -.000

.81

2.81

Ø.50

① JAW

Ø.312 DR.
.375–16 UNF 2B
2 HOLES

2.50
.500
.50
.375
.750
.000

.750
.000
.375
.500
1.50
4.50
90°

Ø .4375 REAM
C'SINK OTHER SIDE
ON THIS HOLE ONLY

NOTE:
1. COUNTERSINK HOLES
2. .030 CHAMFER ALL EDGES

② JAW

.406 DR. 2 HOLES

2.50
.500
.750
.000
.375

1.50
.750
90°
3.00

Precision Machining Technology		LATHE DOG		
UNLESS OTHERWISE SPECIFIED DIMENSIONS ARE IN INCHES INTERPRET DIMENSIONS AND TOLERANCES PER ASME Y14.5M-1994	DESIGNER	JAMES G. HILLWIG JR	MATERIAL	1. C.R.S. 3. C.R.S.
				2. C.R.S. 4. C.R.S.
TOLERANCES .X ±0.032 .XXX ±0.005 .XX ±0.015 ANGLES ± 1 DEG. FRACTIONS ± 1/64	HEAT TREAT	N/N/CARB/CARB		
	NOT TO SCALE			SHEET 1 OF 1

NIMS Skills Practiced

- Job Process Planning
- Layout
- Turning Operations: Chucking
- Drilling and Tapping
- Surface Grinding and Safety
- Vertical Milling and Safety
- Squaring up a Block

Equipment

Power saw, milling machine, lathe, surface grinder

Tools Needed

Scribe	6" steel rule	Layout dye
Dead blow hammer	File	File card
Various drill bits	3/8-16 tap	Stamps
90° countersink	Various sizes of endmills	Parallels
Edge finder	Center drill	Vise stop
.4375 reamer	Grinding wheel	Grinder vise
Tramming items for mill	Adjustable mill angle	Thread triangles
Various types of turning, threading, and facing tools		

Material Needed

- 3/4" square × 4-5/8" long CRS (Jaw Part #1)
- 3/4" square × 3-1/8" long CRS (Jaw Part #2)
- 2 @ 5/8" diameter × 4-1/8" long (Screws Part #3)
- 1/2" diameter × 3" long CRS (Pin Part #4)

Safety

1. Follow all shop safety rules given to you by your instructor and be sure to pass any safety tests given before using shop equipment.
2. Follow all safety rules for power equipment.
3. Be careful of sharp tools when using them.
4. Make sure the file has a handle on it.
5. Be sure all cutting tools are sharp and care is taken while handling them.
6. Be sure ways are locked on mill when necessary.
7. Be sure chuck is tightened and chuck wrench is never left in chuck.
8. Make sure stock is supported with center when necessary.
9. Ring grinding wheels before using them.
10. Make sure magnetic chuck is on when surface grinding.

Order of Operations

MILL

1. Look over blueprint to determine tolerances, tools, and material needed.
2. Gather tools needed, figure out RPMs for required tools, and write down their RPMs on sheet.
3. Measure, mark, and cut all stock needed.
4. Make sure milling machine is in tram.
5. Deburr all over.
6. Place piece in mill vise on parallels with sawed end sticking out; tap down with dead blow hammer to seat part on parallels, then mill just enough off the end to clean it up.
7. Deburr the end you just milled, flip part 180° on parallels in vise, tap down with dead blow hammer to seat part on parallels, and mill to length.
8. Repeat with part #2.
9. Lay out parts.
10. Place part #1 in vise on parallels against vise stop, tap down with dead blow hammer to seat part on parallels, and set vise stop on left side of part.
11. Insert edge finder in collet, set RPMs to 1100, and edge find the X-axis and the Y-axis. Be sure to allow for 1/2 grind stock on the X-axis.
    a. X-axis should read −.1075 (this includes 1/2 of the edge finder and 1/2 of the grind stock).
    b. Y-axis should read .1000 (with the tolerance on the print, the 3/4" stock size should leave plenty of room for grinding).
12. Place center drill in chuck and move to hole locations.
13. Lower center drill to make sure it aligns with layout lines.
14. If location is correct, center drill hole at 1200 RPMs, change to drill, set the RPMs, and drill through part.
15. Repeat with each hole on parts #1 and #2.
16. Countersink holes where indicated on print.
17. You can power tap the two tapped holes in part #1 or use a tap stand.
18. Lay out the lines for the angle and set parts together on an adjustable mill angle and mill the angle.

SURFACE GRINDER

MAKE SURE MAGNETIC CHUCK IS ON WHEN SURFACE GRINDING!

19. Surface grind all sides square and parallel. Deburr after each side is ground.
20. Surface grind the face to clean it up, deburr, then flip the part to grind opposite side.
21. Deburr all edges and place in grinder vise on parallels with ground sides against jaws, leaving one edge sticking out of side of vise so you can turn vise on its side.
22. Grind top of part to clean up and turn vise on side without removing or disturbing part while in vise and grind the end just enough to clean it up.
23. Remove from vise, deburr all edges, place ground side down on chuck with square blocking on each side, and grind bottom to clean it up.
24. Deburr and place last side up with square blocking on each side and clean up surface.
25. Now check sizes and hole location; grind them into location and part to size. Be sure to deburr and block part when needed.
26. Repeat with part #2.
27. Lay flat on grinder chuck and grind a .030-by-45° chamfer on all edges.

LATHE

28. Chuck 5/8" diameter stock for screw in lathe face until end is clean.
29. Mark faced end at .750; this will be the head of the screw.
30. File the convex on the face of the screw.

31. Turn part end for end, face to length, and center drill using a #3 center drill. The center-drilled hole remains in the part for turning and threading, and then it will be removed.

32. Turn .375 diameter back 3-3/8" to the .750 mark of the head.

33. Chase the thread on the screw.

34. Face off center drill and take to length, and then file the radius on the end.

35. Set screw in mill or drill press and drill the .281 hole in the head.

36. Place the 1/2"-diameter stock in the lathe, face both ends, and take to length.

37. Put a radius on end.

38. Flip end for end and turn .438 diameter back .81.

39. Place pin in part #1 through the non-countersunk side.

40. Place the pin radius on a piece of aluminum on top of an anvil, and hammer the pin into the countersink.

41. Place in grinder vise with the mushroomed end sticking up and surface grind until pin and part #1 are blended. (If done correctly, they should look like one piece.)

42. Assemble the rest of the part.

43. Evaluate and turn in.

Student Name: _____ Date Submitted: _____

Class: _____ Total Hours on Job: _____

	Print Dimensions AND Tolerances	Student's Inspection	Instructor's Inspection	Instructor Comments
1	Is the part free of burrs and does it meet finish requirements?			
2	Is the part properly identified?			
3	**{Part #1 - JAW}** 4.50 ± .015			
4	.75 ± .015 (Square)			
5	(Holes) .500 ± .005			
6	(Holes) 2.000 ± .005			
7	(.312 Drill / Tap) .375−16 Threads			
8	90° ± 1° Angle .50 ± .015			
9	90° ± 1° Angle 1.50 ± .015			
10	**{Part #2 - JAW}** 3.00 ± .015			
11	.75 ± .015 (Square)			
12	(Holes) .500 ± .005			
13	(Holes) 2.000 ± .005			
14	(Holes) .406 ± .005			
15	90° ± 1° Angle .50 ± .015			
16	90° ± 1° Angle 1.50 ± .015			
17	Are the holes countersunk?			
18	**{Part #3—Screw 1}** 3.875 ± .005			
19	3.125 ± .005			
20	.375 ± .000 − .005			
21	.375−16 Threads			
22	.625 Diameter .750 ± .005			
23	.281 Hole ± .005			
24	.281 Hole ± .005			
25	**{Part #3—Screw 2}** 3.875 ± .005			

(Continued)

	Print Dimensions AND Tolerances	Student's Inspection	Instructor's Inspection	Instructor Comments
26	3.125 ± .005			
27	.375 ± .000 − .005			
28	.375—16 Threads			
29	.625 Diameter .750 ± .005			
30	.281 Hole ± .005			
31	**{Part #4—Pin}** .50 ± .015			
32	Does the radius look good?			
33	Are there 45°-by-.030 chamfers on the edges of the parts?			
34	Is the pin properly ground to blend into the part?			

PROJECT GRADE: _____

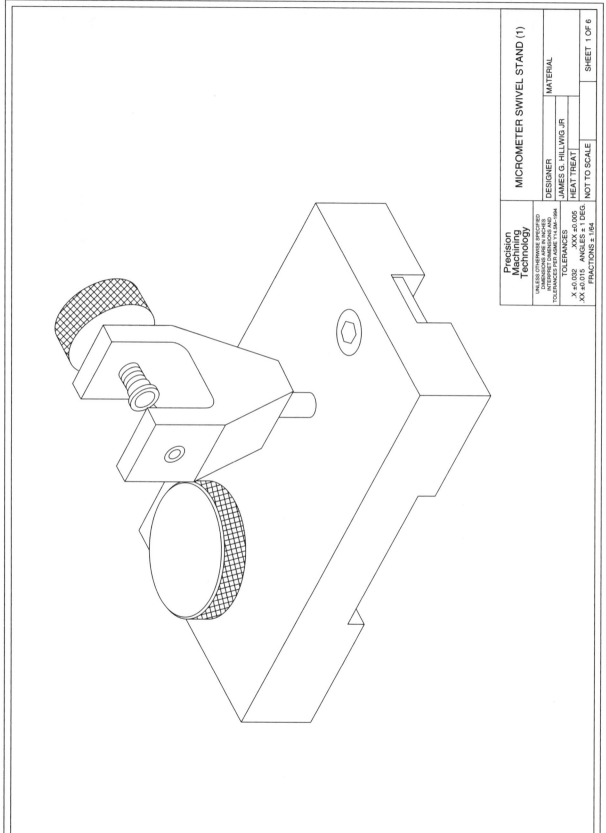

MICROMETER SWIVEL STAND (1)

Precision Machining Technology		MATERIAL	
UNLESS OTHERWISE SPECIFIED DIMENSIONS ARE IN INCHES INTERPRET DIMENSIONS AND TOLERANCES PER ASME Y14.5M–1994	DESIGNER		
	JAMES G. HILLWIG JR		
TOLERANCES	HEAT TREAT		
.X ±0.032 .XXX ±0.005			
.XX ±0.015 ANGLES ± 1 DEG.	NOT TO SCALE		SHEET 1 OF 6
FRACTIONS ± 1/64			

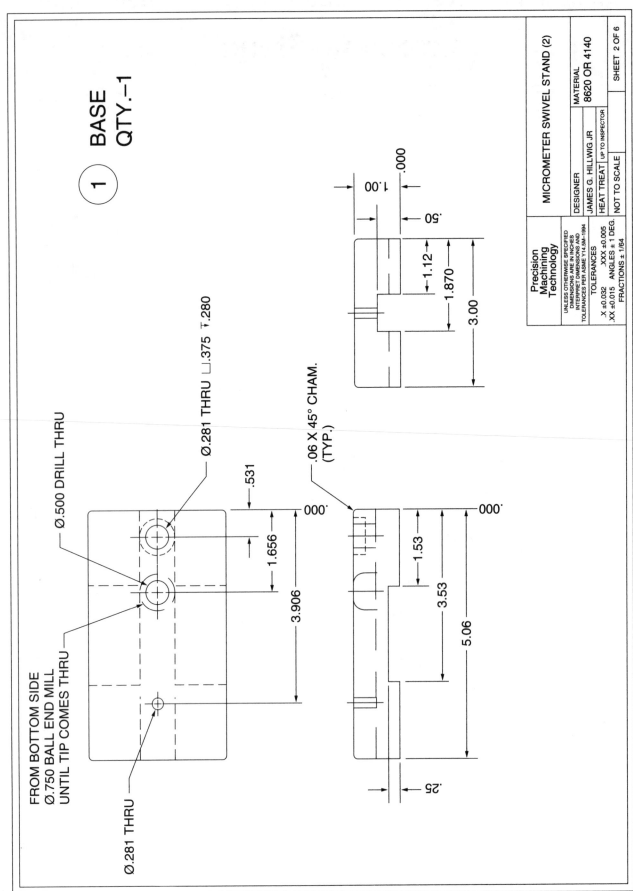

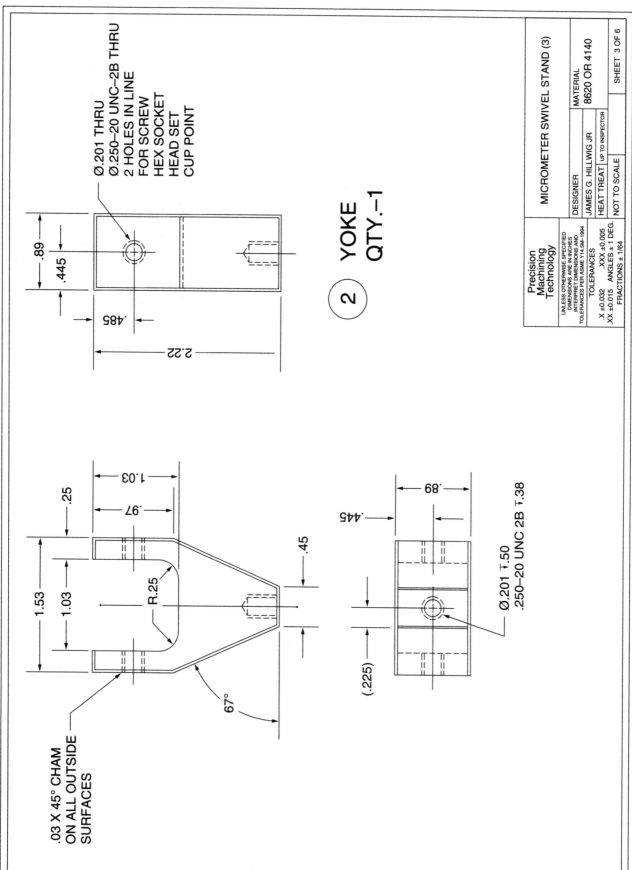

Ø.201 THRU
Ø.250–20 UNC–2B THRU
2 HOLES IN LINE
FOR SCREW
HEX SOCKET
HEAD SET
CUP POINT

.89
.445
.485
2.22

② YOKE
QTY.–1

1.03
.97
.25
1.53
1.03
R.25
67°
.45

.03 X 45° CHAM
ON ALL OUTSIDE
SURFACES

.445
.89
.225
Ø.201 ⊤.50
.250–20 UNC 2B ⊤.38

Precision
Machining
Technology

MICROMETER SWIVEL STAND (3)

DESIGNER
JAMES G. HILLWIG JR

MATERIAL
8620 OR 4140

HEAT TREAT UP TO INSPECTOR

NOT TO SCALE

SHEET 3 OF 6

UNLESS OTHERWISE SPECIFIED
DIMENSIONS ARE IN INCHES
INTERPRET DIMENSIONS AND
TOLERANCES PER ASME Y14.5M–1994
TOLERANCES
.X ±0.032   .XXX ±0.005
.XX ±0.015   ANGLES ± 1 DEG.
FRACTIONS ± 1/64

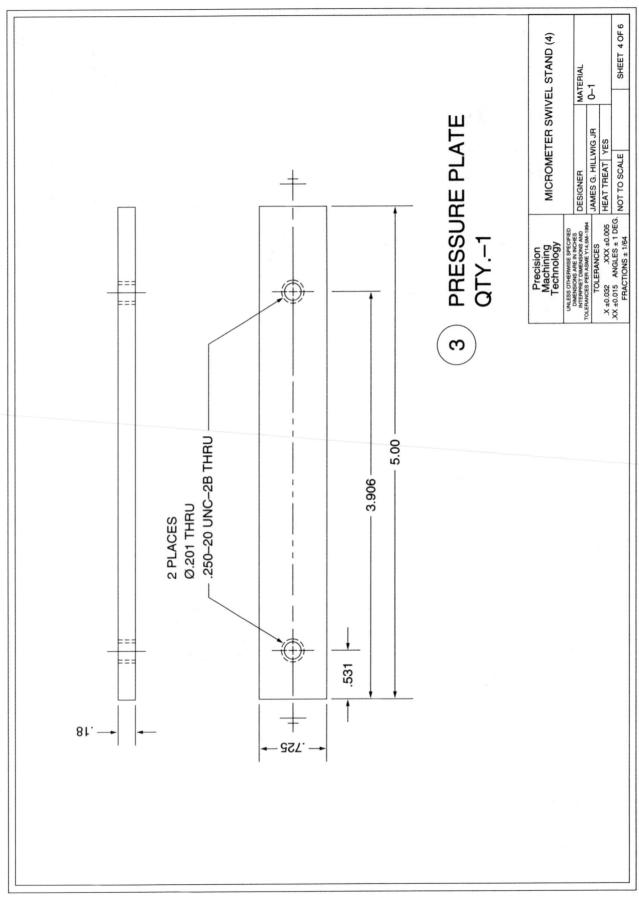

2 PLACES
Ø.201 THRU
.250–20 UNC–2B THRU

5.00

3.906

.531

.725

.18

3 PRESSURE PLATE
QTY.–1

Precision Machining Technology		MICROMETER SWIVEL STAND (4)	
UNLESS OTHERWISE SPECIFIED DIMENSIONS ARE IN INCHES INTERPRET DIMENSIONS AND TOLERANCES PER ASME Y14.5M-1994		DESIGNER	MATERIAL
		JAMES G. HILLWIG JR	O–1
TOLERANCES .X ±0.032 .XXX ±0.005 .XX ±0.015 ANGLES ± 1 DEG. FRACTIONS ± 1/64		HEAT TREAT YES	
		NOT TO SCALE	SHEET 4 OF 6

©Cengage Learning 2012

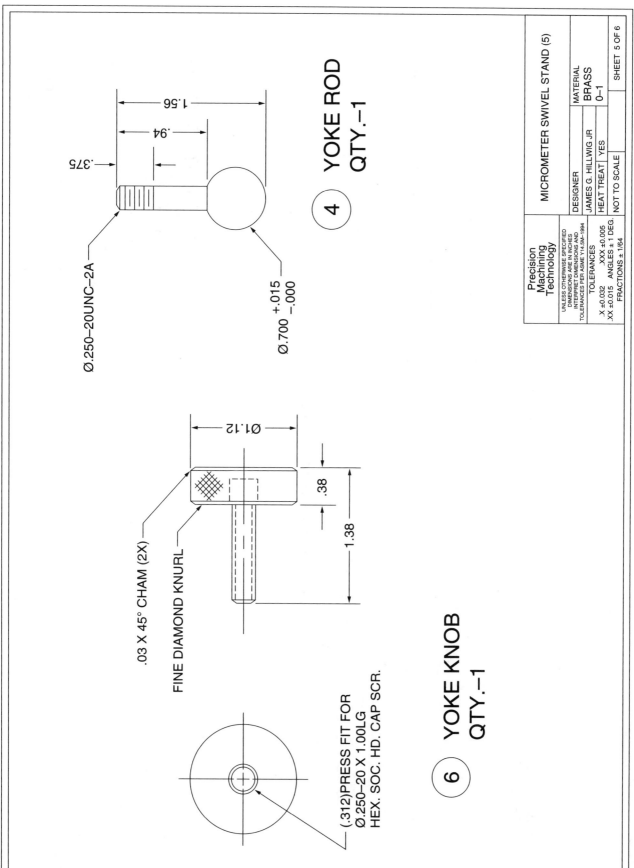

Ø.250–20UNC–2A

.375

.94

1.56

Ø.700 +.015 −.000

④ YOKE ROD QTY.–1

.03 X 45° CHAM (2X)

FINE DIAMOND KNURL

Ø1.12

.38

1.38

(.312)PRESS FIT FOR Ø.250–20 X 1.00LG HEX. SOC. HD. CAP SCR.

⑥ YOKE KNOB QTY.–1

Precision Machining Technology	MICROMETER SWIVEL STAND (5)	
	DESIGNER	MATERIAL
UNLESS OTHERWISE SPECIFIED DIMENSIONS ARE IN INCHES INTERPRET DIMENSIONS AND TOLERANCES PER ASME Y14.5M–1994	JAMES G. HILLWIG JR	BRASS
	HEAT TREAT YES	0–1
TOLERANCES .X ±0.032 .XXX ±0.005 .XX ±0.015 ANGLES ± 1 DEG. FRACTIONS ± 1/64	NOT TO SCALE	SHEET 5 OF 6

.03 X 45° CHAM (2X)

FINE DIAMOND KNURL

Ø1.5

.34

.000

.64

.375

TO ACCEPT A Ø.250–20 UNC x 1.25 LG SET SCREW

Ø.201 ⊤.550
Ø.250–20 UNC 2B ⊤.450

⑤ ADJUSTING KNOB
QTY.–1
BRASS

Precision Machining Technology	MICROMETER SWIVEL STAND (6)		
UNLESS OTHERWISE SPECIFIED DIMENSIONS ARE IN INCHES INTERPRET DIMENSIONS AND TOLERANCES PER ASME Y14.5M–1994	DESIGNER	MATERIAL	
	JAMES G. HILLWIG JR	BRASS	
TOLERANCES	HEAT TREAT		
X ±0.032 XXX ±0.005			
XX ±0.015 ANGLES ± 1 DEG.	NOT TO SCALE	SHEET 6 OF 6	
FRACTIONS ± 1/64			

⑤

NIMS Skills Practiced

- Job Process Planning
- Layout
- Turning Operations: Chucking
- Drilling and Tapping
- Surface Grinding and Safety
- Vertical Milling and Safety
- Squaring up a Block
- CNC Milling
- CNC Turning

Equipment

Power saw, milling machine, lathe, CNC mill, CNC lathe, surface grinder

Tools Needed

Scribe	6" steel rule	Layout dye
Dead blow hammer	File	File card
Various drill bits	1/4-20 tap	Stamps
90° countersink	Various sizes of endmills	Parallels
Edge finder	Center drill	Vise stop
Grinding wheel	Grinder vise	Tramming items for mill
Adjustable mill angle	Thread triangles	Various types of turning, threading, and facing tools

Material Needed

- 1" × 3" × 5.185" long CRS (Base Part #1)
- 1-1/8" × 1-3/4" × 2-3/8" long CRS (Yoke Part #2)
- .1875 × .750 × 5-1/8" long 0-1 (Pressure Plate Part #3)
- 3/4" diameter × 1-3/4" long 0-1 (Yoke Knob Part #4)
- 1.125 diameter × .38 brass (Yoke Knob Part #6—use enough stock to safely hold part)
- 1.5 diameter × .64 brass (Adjusting Knob Part #5—use enough stock to safely hold part)

Safety

1. Follow all shop safety rules given to you by your instructor and be sure to pass any safety tests given before using shop equipment.
2. Follow all safety rules for power equipment.
3. Be careful of sharp tools when using them.
4. Make sure the file has a handle on it.
5. Be sure all cutting tools are sharp and care is taken while handling them.
6. Be sure ways are locked on mill when necessary.
7. Be sure chuck is tightened and chuck wrench is never left in chuck.
8. Make sure stock is supported with center when necessary.
9. Ring grinding wheels before using them.
10. Make sure magnetic chuck is on when surface grinding.
11. Follow all CNC rules and safety guidelines for operation and programming.

Order of Operations

BASE

MILL

1. Look over blueprint to determine tolerances, tools, and material needed.
2. Gather tools needed, figure out RPMs for required tools, and write down their RPMs on sheet.
3. Measure, mark, and cut all stock needed.
4. Make sure milling machine is in tram.
5. Deburr all over.
6. Place piece in mill vise on parallels, square it up, and take to size.
7. Deburr part.
8. Lay out part.
9. Edge find part.
10. Center drill and drill holes.
11. From top of part, counterbore hole where indicated.
12. On bottom of part, using a 3/4" ball endmill, drill in .500 hole until tip of endmill breaks through .500 hole. This will create the radius for the ball to rotate on.
13. With the part still in vise, mill out the two slots.
14. Countersink holes where indicated on print.
15. Send for heat treatment.

YOKE

CNC MILL

1. Following your instructor's directions, cut stock to size and write the program.
2. Deburr the part and place in the CNC mill.
3. Load program and tools and follow your instructor's directions to cut out part.
4. You can drill and tap 1/4-20 holes using a manual mill or program the CNC.
5. Send for heat treatment.

PRESSURE PLATE

MILL

1. Cut, deburr, square, and mill part to size.
2. Drill and tap both holes for 1/4-20 screws.
3. Send for heat treatment.

YOKE ROD

CNC LATHE

1. Following your instructor's directions, cut stock to size recommended by your instructor and write the program.
2. Deburr the part and place in the CNC lathe.
3. Load program and tools and follow your instructor's directions to cut out part.
4. Send for heat treatment.

YOKE KNOB

CNC OR MANUAL LATHE

1. If using the CNC lathe, follow steps above.
2. If using manual lathe, cut a bar of brass approximately 12" long (so you can safely hold in chuck), face one end, center drill, and drill to print dimensions.
3. Slide a few inches out of chuck and knurl.
4. Cut off part. Partway through cutoff, stop. Using a file, file chamfers on both sides of knob. Continue cutting off part.
5. Press screw into knob and polish top.

ADJUSTING KNOB

CNC OR MANUAL LATHE

1. If using the CNC lathe, follow steps above.
2. If using manual lathe, cut a bar of brass approximately 12" long (so you can safely hold in chuck), face one end, center drill, and drill to print dimensions.
3. Slide a few inches out of chuck and knurl.
4. Turn down .375 shoulder.
5. Following your instructor's directions, you can tap the hole on the lathe.
6. Cut off part. Partway through cutoff, stop. Using a file, file chamfers on both sides of knob. Continue cutting off part.
7. Loctite screw into knob and polish top.

SURFACE GRINDER

MAKE SURE MAGNETIC CHUCK IS ON WHEN SURFACE GRINDING!

1. Surface grind all sides square and parallel on each part. Deburr after each side is ground.
2. Lay flat on grinder chuck and grind chamfer on all edges where indicated on print.

The part looks really nice if it is sent out for blackening when all of the grinding is complete. On the two 1/4-20 screws that go into the top of the yoke, if you get a couple of black rubber-flanged sink washers, run a tap through the hole in the washers and thread them on the inside of the screws. They will clamp your micrometers without the screws marring the surface of the micrometers.

Student Name: _____ Date Submitted: _____

Class: _____ Total Hours on Job: _____

	Print Dimensions AND Tolerances	Student's Inspection	Instructor's Inspection	Instructor Comments
1	Is the part free of burrs and does it meet finish requirements?			
2	Is the part properly identified?			
3	Base Length 5.06 ± .015			
4	Base Width 3.00 ± .015			
5	Base Height 1.00 ± .015			
6	Hole Length 3.375 ± .015			
7	Hole Length 1.125 ± .015			
8	Hole Location .531 ± .005			
9	Base Slot 2.00 ± .015			
10	Base Slot .75 ± .015			
11	Yoke Length 2.22 ± .015			
12	Yoke Width 1.53 ± .015			
13	Yoke Height .89 ± .015			
14	Yoke Opening 1.03 ± .015			
15	Yoke Arms .25 ± .015			
16	Yoke Angle 67° ± 1/2°			
17	Pressure Plate Length 5.00 ± .015			
18	Pressure Plate Width .725 ± .005			
19	Pressure Plate Height .18 ± .015			
20	Hole Length 3.375 ± .005			

PROJECT GRADE: _____

Tapping Block

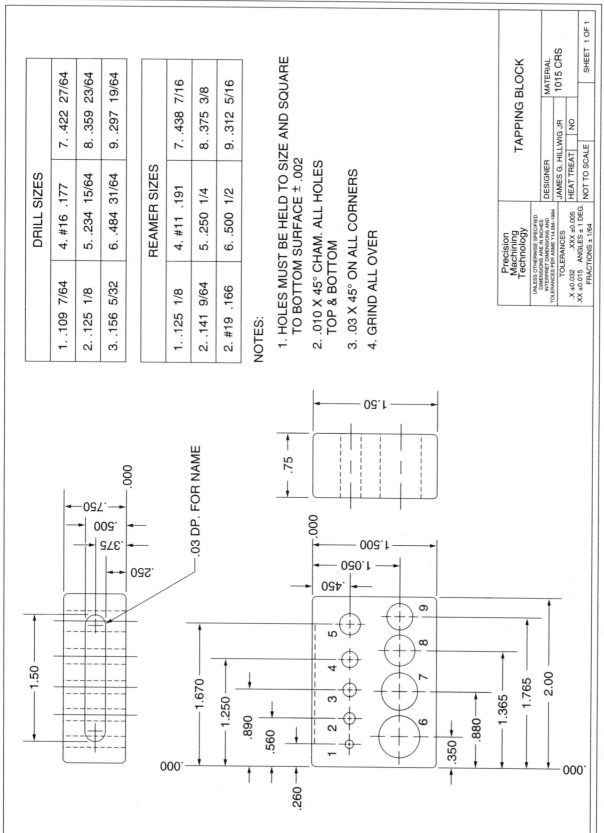

DRILL SIZES

1. .109 7/64	4. #16 .177	7. .422 27/64
2. .125 1/8	5. .234 15/64	8. .359 23/64
3. .156 5/32	6. .484 31/64	9. .297 19/64

REAMER SIZES

1. .125 1/8	4. #11 .191	7. .438 7/16
2. .141 9/64	5. .250 1/4	8. .375 3/8
2. #19 .166	6. .500 1/2	9. .312 5/16

NOTES:

1. HOLES MUST BE HELD TO SIZE AND SQUARE TO BOTTOM SURFACE ± .002

2. .010 X 45° CHAM. ALL HOLES TOP & BOTTOM

3. .03 X 45° ON ALL CORNERS

4. GRIND ALL OVER

.03 DP. FOR NAME

Precision Machining Technology

UNLESS OTHERWISE SPECIFIED DIMENSIONS ARE IN INCHES INTERPRET DIMENSIONS AND TOLERANCES PER ASME Y14.5M–1994

TOLERANCES
.X ±0.032   .XXX ±0.005
.XX ±0.015   ANGLES ± 1 DEG.
FRACTIONS ± 1/64

DESIGNER
JAMES G. HILLWIG JR
HEAT TREAT   NO
NOT TO SCALE

TAPPING BLOCK

MATERIAL
1015 CRS

SHEET 1 OF 1

©Cengage Learning 2012

NIMS Skills Practiced

- Job Process Planning
- Layout
- Surface Grinding and Safety
- Vertical Milling and Safety
- Squaring up a Block

Equipment

Power saw, milling machine, surface grinder

Tools Needed

Scribe	6" steel rule	Layout dye
Hammer	File	File card
Various sizes of drill bits	Various sizes of reamers	Stamps
90° countersink	Various sizes of endmills	Parallels
Edge finder	Center drills	Vise stop
Machinist square	Grinding wheel	Grinder vise
Dead blow hammer	Tramming items for mill	

Material Needed

- 2.1 × 1.5 × .75 mild steel

Safety

1. Follow all shop safety rules given to you by your instructor and be sure to pass any safety tests given before using shop equipment.
2. Follow all safety rules for power equipment.
3. Be careful of sharp tools when using them.
4. Make sure the file has a handle on it.
5. Be sure all cutting tools are sharp and care is taken while handling them.
6. Be sure ways are locked on mill when necessary.
7. Ring grinding wheels before using them.
8. Make sure magnetic chuck is on when surface grinding.

Order of Operations

1. Look over blueprint to determine tolerances, tools, and material needed.
2. Gather tools needed, figure out RPMs for required tools, and write down their RPMs on sheet.
3. Measure, mark, and cut stock 3/4" thick × 1 1/2" wide; saw 2-1/8" long.
4. Make sure milling machine is in tram.
5. Deburr all over; check squareness on rolled sides from steel mill.
6. Place piece in mill vise with sawed end up, placing machinist square next to part, and mill just enough off the end to clean it up.
7. Deburr the end you just milled, place the milled end on the bottom of the vise, turn or rotate piece 90° keeping the milled end on the bottom of the vise, and use the machinist square again to mill just enough off to clean this end up.

8. Deburr the end you just milled and place in the bottom of vise and take a light cut off the part; this will be the first end you cut.

9. If you did this correctly, the ends will be square—be sure to check them.

10. Mill part 2.015 long, deburr, and re-check for squareness.

11. Paint part with layout dye, then use height gage to lay out hole locations, being sure to leave grind stock on each end.

12. Place part in vise on parallels against vise stop; tap down with dead blow hammer to seat part on parallels.

13. Insert edge finder in collet, set RPMs to 1,100, and edge find the X-axis and the Y-axis. Be sure to allow for 1/2 grind stock on the X-axis.

    a. X-axis should read −.1075 (this includes 1/2 of the grind stock).

    b. Y-axis should read .1000.

14. Place center drill in chuck and move to −.450 on Y-axis and .260 on X-axis.

15. Lower center drill to make sure it aligns with layout lines.

16. If location is correct, center drill hole at 1,200 RPMs, change to 7/64 drill, set the RPMs, and drill through part; then insert 1/8 reamer, set RPMs, and ream hole through.

17. Move X-axis to .560, put center drill back in, lower to part, and check location to layout lines; if correct, set RPMs and center drill hole. Change to 7/64 drill, set the RPMs, and drill through part—this is called step drilling; then insert 1/8 drill, set RPMs, and drill through part. Insert 9/64 reamer, set RPMs, and ream through hole.

18. Repeat this process for each hole—step drilling each hole, drilling hole size, then reaming hole.

19. Remove part from vise and deburr holes.

20. Return part to vise against vise stop and countersink each hole, both sides. Remember to set the RPMs for the countersink.

21. Lay out name pocket, 3/8 wide × 1" long × .030 deep, centered on part.

22. Using a 3/8 endmill, set RPMs, locate pocket, and mill to print dimensions; remove part, deburr, place on anvil, and, using steel stamps, stamp your name in pocket.

MAKE SURE MAGNETIC CHUCK IS ON WHEN SURFACE GRINDING!

23. Surface grind all sides square and parallel. Deburr after each side is ground.

24. Surface grind the face to clean it up, deburr, then flip the part to grind bottom.

25. Deburr all edges, and place all the way down in grinder vise with ground sides against jaws, leaving edge with name pocket sticking out of top of vise. Leave one edge sticking out of side of vise, so you can turn vise on its side.

26. Grind top of part to clean up and turn vise on side without removing or disturbing part while in vise just to clean it up.

27. Remove from vise, deburr all edges, and place name pocket side down on chuck with square blocking on each side and grind side to clean it up.

28. Deburr and place last side up with square blocking on each side and clean up surface.

29. Now check hole locations and grind them into location and part to size. Be sure to deburr and block part when needed.

30. Grind a .030 × 45° chamfer on all edges.

31. Inspect the part to print dimensions; record on inspection sheet and turn in.

How to Determine the RPMs

Tool Being Used	SFPM	Formula	RPM
Edge Finder	80	Given	1,000
#2 Center Drill	80	Given	1,200
#3 Center Drill	80	Given	1,100
Drill 7/6	80	$\dfrac{4 \times 80}{.109}$	
Drill 1/8	80	$\dfrac{4 \times 80}{.125}$	
Drill 5/32	80	$\dfrac{4 \times 80}{.156}$	
Drill #16	80	$\dfrac{4 \times 80}{.177}$	
Drill 15/64	80	$\dfrac{4 \times 80}{.234}$	
Drill 31/64	80	$\dfrac{4 \times 80}{.484}$	
Drill 27/64	80	$\dfrac{4 \times 80}{.422}$	
Drill 23/64	80	$\dfrac{4 \times 80}{.359}$	
Drill 19/64	80	$\dfrac{4 \times 80}{.297}$	

How to Determine the RPMs for Reamers

Tool Being Used	SFPM	Formula	RPM
Reamer 1/8	80	$\dfrac{4 \times 80}{.125}$	Divide RPM in half for reamers
Reamer 9/64	80	$\dfrac{4 \times 80}{.141}$	Divide RPM in half for reamers
Reamer #19	80	$\dfrac{4 \times 80}{.166}$	Divide RPM in half for reamers
Reamer #11	80	$\dfrac{4 \times 80}{.191}$	Divide RPM in half for reamers
Reamer 1/4	80	$\dfrac{4 \times 80}{.250}$	Divide RPM in half for reamers
Reamer 1/2	80	$\dfrac{4 \times 80}{.500}$	Divide RPM in half for reamers
Reamer 7/16	80	$\dfrac{4 \times 80}{.438}$	Divide RPM in half for reamers
Reamer 3/8	80	$\dfrac{4 \times 80}{.375}$	Divide RPM in half for reamers
Reamer 5/16	80	$\dfrac{4 \times 80}{.312}$	Divide RPM in half for reamers
90° Countersink	80	$\dfrac{4 \times 80}{\text{Tool } \varnothing}$	

Student Name: _____   Date Submitted: _____

Class: _____   Total Hours on Job: _____

	Print Dimensions AND Tolerances	Student's Inspection	Instructor's Inspection	Instructor Comments
1	Is the part free of burrs and does it meet finish requirements?			
2	Is the part properly identified?			
3	**{Top View}** 2.00 ± .015			
4	**{Side View}** 1.50 ± .015			
5	**{Top View}** .75 ± .015			
6	**{Front View}** Hole #1 1/8" Reamer ± .015			
7	**{Front View}** Hole #1 .260 ± .005 by .450 ± .005			
	Top Row of Holes			
8	**{Front View}** Holes #2 through #5 Reamed to Dimensions ± .015			
9	**{Front View}** Hole #2 .560 ± .005 by .450 ± .005			
10	**{Front View}** Hole #3 .890 ± .005 by .450 ± .005			
11	**{Front View}** Hole #4 1.250 ± .005 by .450 ± .005			
12	**{Front View}** Hole #5 1.670 ± .005 by .450 ± .005			
	Bottom Row of Holes			
13	**{Front View}** Holes #6 through #9 Reamed to Dimensions ± .015			
14	**{Front View}** Hole #6 .350 ± .005 by .450 ± .005			
15	**{Front View}** Hole #7 .880 ± .005 by .450 ± .005			
16	**{Front View}** Hole #8 1.365 ± .005 by .450 ± .005			
17	**{Front View}** Hole #9 1.765 ± .005 by .450 ± .005			
18	Grind Finish			
19	Chamfers on Outside of Part and Holes			

PROJECT GRADE: _____

Bench Block

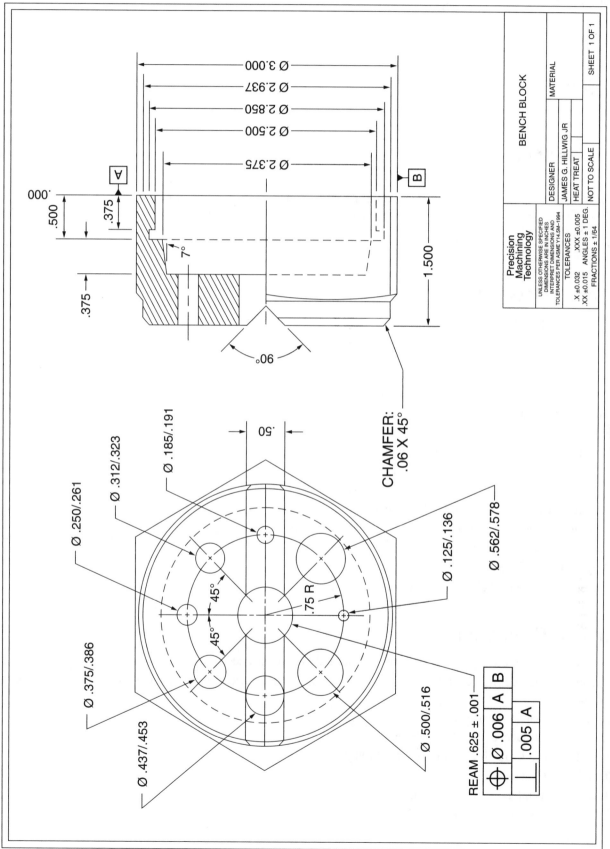

CHAMFER:
.06 X 45°

Ø 3.000
Ø 2.937
Ø 2.850
Ø 2.500
Ø 2.375

.000
.500
.375
.375

7°

90°

1.500

.50

Ø .185/.191
Ø .312/.323
Ø .250/.261

45°
45°
.75 R

Ø .375/.386
Ø .437/.453
Ø .500/.516

Ø .125/.136
Ø .562/.578

REAM .625 ± .001

⊕ | Ø .006 | A | B
⊥ | .005 | A

BENCH BLOCK

Precision
Machining
Technology

UNLESS OTHERWISE SPECIFIED
DIMENSIONS ARE IN INCHES
INTERPRET DIMENSIONS AND
TOLERANCES PER ASME Y14.5M–1994
TOLERANCES
.X ±0.032   .XXX ±0.005
.XX ±0.015   ANGLES ± 1 DEG.
FRACTIONS ± 1/64

DESIGNER
JAMES G. HILLWIG JR

MATERIAL

HEAT TREAT

NOT TO SCALE

SHEET 1 OF 1

NIMS Skills Practiced

- Job Process Planning
- Turning Operations: Chucking
- Milling
- Drilling
- Grinding

Equipment

Power saw, lathe, bench grinder, mill and surface grinders

Tools Needed

Marker
Various drills
Hammer
Various types of turning, facing,
   and boring tools

6" steel rule
Adjustable mill angle
Layout dye
5/8" reamer

Center drills
Center punch
Dividers

Material Needed

- Hex stock (CRS): 3" hex × 1-5/8" long

Safety

1. Follow all shop safety rules given to you by your instructor and be sure to pass any safety tests given before using shop equipment.
2. Follow all safety rules for power equipment.
3. Be careful of sharp tools when using them.
4. Make sure the files have a handle on them.
5. Be sure all cutting tools are sharp and care is taken while handling them.
6. Be sure ways are locked on mill when necessary.
7. Be sure chuck is tightened and chuck wrench is never left in chuck.
8. Make sure stock is supported with center when necessary.
9. Ring grinding wheels before using them.
10. Make sure magnetic chuck is on when surface grinding.

Order of Operations

1. Look over blueprint to determine tolerances, tools, and material needed.
2. Gather tools needed, figure out RPMs for required stock diameters and drills used, and write down their RPMs on sheet.
3. Measure and cut stock sizes needed.
4. Chuck in lathe, face, and center drill top.
5. Step drill the hole out to 19/32, then ream through with a 5/8" reamer.
6. Turn 2-15/16" diameter back 3/8".
7. Turn a 45° chamfer × 1/16" on front.
8. Turn end for end in lathe and step drill as close to 2" as you can.

9. Remove the rest with a boring tool to 2" inside diameter 7/8" deep.

10. Bore 2-1/2" diameter 1/2" in from end.

11. Now with the boring tool, turn the inside angle of 7° in the bottom 3/8" of hole until the large diameter is 2-3/8".

12. Lay out to holes on the top to be drilled, center punch the holes, and drill them on the drill press—first with a center drill then the step drill or drill if needed and then to size.

13. Lay out the 45° angle marks on the top, place part in mill vise on adjustable mill angle, and mill the angle.

14. Mill name pocket in center of one of the flats and stamp name in it.

15. Deburr part and place on surface grinder and grind top just to clean it up.

16. Evaluate and turn in.

Student Name: _____ Date Submitted: _____

Class: _____ Total Hours on Job: _____

	Print Dimensions AND Tolerances	Student's Inspection	Instructor's Inspection	Instructor Comments
1	Is the part free of burrs and does it meet finish requirements?			
2	Is the part properly identified?			
3	Height 1-1/2" ± .015			
4	Hole 1/4" ± .015			
5	Hole 5/16" ± .015			
6	Hole 3/16" ± .015			
7	Hole 9/16" ± .015			
8	Hole 1/8" ± .015			
9	Hole 1/2" ± .015			
10	Hole 7/16" ± .015			
11	Hole 3/8" ± .015			
12	Hole 5/8" ± .015			
13	90° Angle 7/16" Wide ± .015			
14	Holes on a 7/8" Radius			
15	Holes 45° Apart			
16	Top Step 2-15/16" ± .015			
17	Top Chamfer 1/16"			
18	Inside Diameter 2-1/2" ± .015			
19	Large Diameter of 7° Taper 2-3/8" ± .015			
20	Depth of Inside 7/8" ± .015			

PROJECT GRADE: _____

Short Diamond Holder

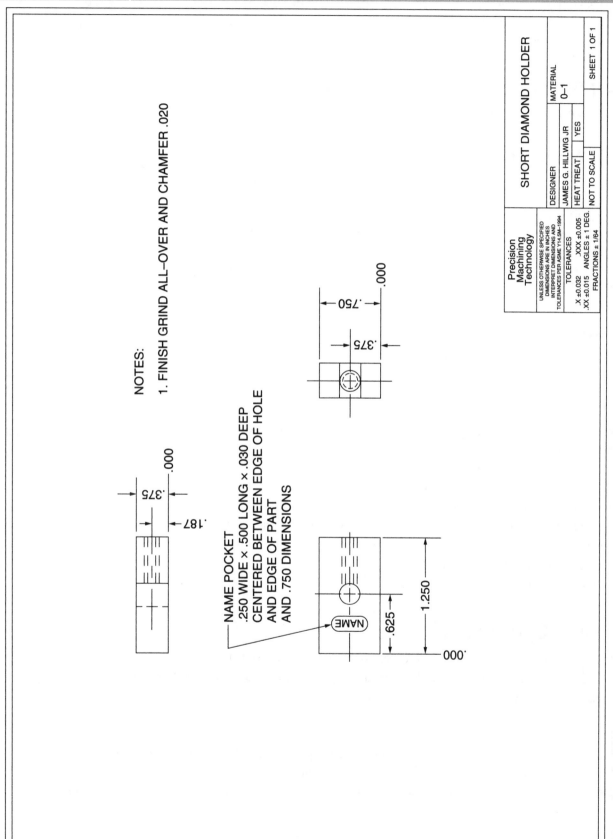

NOTES:

1. FINISH GRIND ALL-OVER AND CHAMFER .020

NAME POCKET
.250 WIDE × .500 LONG × .030 DEEP
CENTERED BETWEEN EDGE OF HOLE
AND EDGE OF PART
AND .750 DIMENSIONS

Precision Machining Technology		SHORT DIAMOND HOLDER	
UNLESS OTHERWISE SPECIFIED DIMENSIONS ARE IN INCHES INTERPRET DIMENSIONS AND TOLERANCES PER ASME Y14.5M–1994		DESIGNER	MATERIAL
TOLERANCES .X ±0.032 .XXX ±0.005 .XX ±0.015 ANGLES ± 1 DEG. FRACTIONS ± 1/64		JAMES G. HILLWIG JR	0–1
		HEAT TREAT YES	
		NOT TO SCALE	SHEET 1 OF 1

.750

.375

.000

.375

.187

.000

1.250

.625

.000

NAME

NIMS Skills Practiced

- Job Process Planning
- Layout
- Surface Grinding and Safety
- Vertical Milling and Safety
- Squaring up a Block

Equipment

Power saw, milling machine, surface grinder

Tools Needed

Scribe	6" steel rule	Layout dye
Dead blow hammer	File	File card
15/64 and #21 drill bits	1/4" reamer	10-32 tap
90° countersink	Various sizes of endmills	Parallels
Edge finder	Center drill	Vise stop
Machinist square	Grinding wheel	Grinder vise
Tramming items for mill	Stamps	Tap stand

Material Needed

- 3/8" × 3/4" × 1-3/8" tool steel (oversize stock)

Safety

1. Follow all shop safety rules given to you by your instructor and be sure to pass any safety tests given before using shop equipment.
2. Follow all safety rules for power equipment.
3. Be careful of sharp tools when using them.
4. Make sure the file has a handle on it.
5. Be sure all cutting tools are sharp and care is taken while handling them.
6. Be sure ways are locked on mill when necessary.
7. Ring grinding wheels before using them.
8. Make sure magnetic chuck is on when surface grinding.

Order of Operations

1. Look over blueprint to determine tolerances, tools, and material needed.
2. Gather tools needed, figure out RPMs for required tools, and write down their RPMs on sheet.
3. Measure, mark, and cut stock 3/8" thick × 3/4" wide; saw 1-3/8" long.
4. Make sure milling machine is in tram.
5. Deburr all over.
6. Place piece in mill vise on parallels with sawed end sticking out, tap down with dead blow hammer to seat part on parallels, and mill just enough off the end to clean it up.
7. Deburr the end you just milled, flip part 180° on parallels in vise, tap down with dead blow hammer to seat part on parallels, and mill to length.

8. Paint part with layout dye and use height gage to lay out hole locations, being sure to leave grind stock on each end.

9. Place part in vise on parallels against vise stop; tap down with dead blow hammer to seat part on parallels, then set vise stop.

10. Insert edge finder in collet, set RPMs to 1,100, and edge find the X-axis and the Y-axis. Be sure to allow for 1/2 grind stock on the X-axis.

   a. X-axis should read −.1075 (this includes 1/2 of the grind stock).

   b. Y-axis should read .1000.

11. Place center drill in chuck and move to −.375 on Y-axis and .625 on X-axis.

12. Lower center drill to make sure it aligns with layout lines.

13. If location is correct, center drill hole at 1,200 RPMs, change to 15/64 drill, set the RPMs and drill through part; then insert 1/4 reamer, set RPMs, and ream hole through.

14. Remove part and deburr, rotate part 90°, and place the 3/4" side against back jaw of vise and 3/8" edge against vise stop.

15. Place center drill in chuck and move to −.187 on Y-axis and .375 on X-axis.

16. Lower center drill to make sure it aligns with layout lines.

17. If location is correct, center drill hole at 1,200 RPMs, change to #21 drill, set the RPMs, and drill −.850 deep; then insert 90° countersink, set RPMs, and countersink hole.

18. Remove part from the vise and deburr hole.

19. Lay out name pocket, 1/4" wide × 1/2" long × .030 deep, centered on 3/4" face centered from edge to edge of hole—this is opposite side of tapped hole.

20. Using a 1/4" endmill, set RPMs, locate pocket, and mill to print dimensions. Remove part, deburr, place on anvil, and using steel stamps, stamp your name in pocket.

21. Using tap stand, tap the 10-32 hole.

MAKE SURE MAGNETIC CHUCK IS ON WHEN SURFACE GRINDING!

22. Surface grind all sides square and parallel. Deburr after each side is ground.

23. Surface grind the face to clean it up, deburr, then flip the part to grind bottom.

24. Deburr all edges and place all the way down in grinder vise with ground sides against jaws, leaving 3/4" edge out of top of vise. Leave 1-3/8" edge sticking out of side of vise so you can turn vise on its side.

25. Grind top of part to clean up and turn vise on side without removing or disturbing part while in vise just to clean it up.

26. Remove from vise, deburr all edges, and place ground side down on chuck with square blocking on each side; grind side to clean it up.

27. Deburr and place last side up with square blocking on each side and clean up surface.

28. Now check hole locations and grind them into location and part to size. Be sure to deburr and block part when needed.

29. Grind a .020 × 45° chamfer on all edges.

30. Inspect the part to print dimensions; record on inspection sheet and turn in.

How to Determine the RPMs

Tool Being Used	Surface Feet per Minute	Formula	RPM
Edge Finder	50	Given	1,000
#2 Center Drill	50	Given	1,200
#3 Center Drill	50	Given	1,100
Drill 15/64	50	$\dfrac{4 \times 50}{.235}$	
Drill #21	50	$\dfrac{4 \times 50}{.125}$	

How to Determine the RPMs for Reamers

Tool Being Used	Surface Feet per Minute	Formula	RPM
Reamer 1/8	50	$\dfrac{4 \times 50}{.250}$	Divide RPM in half for reamers
90° Countersink	50	$\dfrac{4 \times 50}{\text{Tool } \varnothing}$	

Student Name: _____ Date Submitted: _____

Class: _____ Total Hours on Job: _____

	Print Dimensions AND Tolerances	Student's Inspection	Instructor's Inspection	Instructor Comments
1	Is the part free of burrs and does it meet finish requirements?			
2	Is the part properly identified?			
3	.375 ± .005			
4	.750 ± .005			
5	1.250 ± .005			
6	1/4 Hole Location .625 ± .005			
7	1/4 Hole Location .375 ± .005			
8	10-32 Tapped Hole .187 ± .005			
9	10-32 Tapped Hole .375 ± .005			
10	Are the hole chamfers to size?			
11				
12				
13				
14				
15				

PROJECT GRADE: _____

Parallels

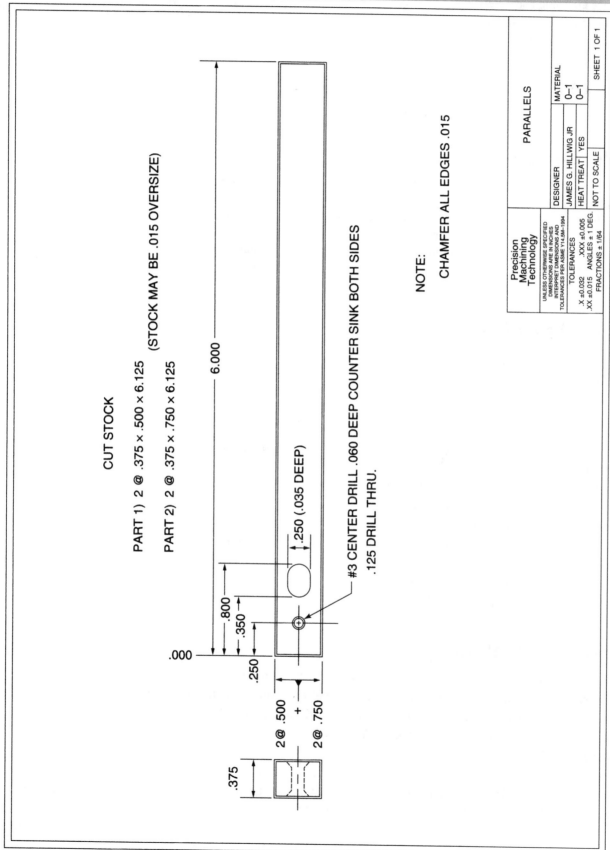

CUT STOCK

PART 1) 2 @ .375 × .500 × 6.125   (STOCK MAY BE .015 OVERSIZE)

PART 2) 2 @ .375 × .750 × 6.125

6.000

.250 (.035 DEEP)

#3 CENTER DRILL .060 DEEP COUNTER SINK BOTH SIDES
.125 DRILL THRU.

.800

.350

.000

.250

.375

2 @ .500

2 @ .750

NOTE:

CHAMFER ALL EDGES .015

Precision Machining Technology	PARALLELS		
UNLESS OTHERWISE SPECIFIED DIMENSIONS ARE IN INCHES INTERPRET DIMENSIONS AND TOLERANCES PER ASME Y14.5M-1994	DESIGNER		MATERIAL
	JAMES G. HILLWIG JR		O—1
TOLERANCES .X ±0.032 .XXX ±0.005 .XX ±0.015 ANGLES ± 1 DEG. FRACTIONS ± 1/64	HEAT TREAT	YES	O—1
	NOT TO SCALE		SHEET 1 OF 1

NIMS Skills Practiced

- Job Process Planning
- Layout
- Surface Grinding and Safety
- Vertical Milling and Safety
- Squaring up a Block

Equipment

Power saw, milling machine, surface grinder

Tools Needed

Scribe
Dead blow hammer
1/8" drill bit
Edge finder
Grinding wheel
Tramming items for mill
Compound clamps

6" steel rule
File
Various sizes of endmills
Center drill
Grinder vise
Large angle plate

Layout dye
File card
Parallels
Vise stop
Stamps
Shim stock

Material Needed

- 2 @ 3/8" × 1/2" × 6-1/8" tool steel (oversize stock)
- 2 @ 3/8" × 3/4" × 6-1/8" tool steel (oversize stock)

Safety

1. Follow all shop safety rules given to you by your instructor and be sure to pass any safety tests given before using shop equipment.
2. Follow all safety rules for power equipment.
3. Be careful of sharp tools when using them.
4. Make sure the file has a handle on it.
5. Be sure all cutting tools are sharp and care is taken while handling them.
6. Be sure ways are locked on mill when necessary.
7. Ring grinding wheels before using them.
8. Make sure magnetic chuck is on when surface grinding.

Order of Operations

1. Look over blueprint to determine tolerances, tools, and material needed.
2. Gather tools needed, figure out RPMs for required tools, and write down their RPMs on sheet.
3. Measure, mark, and cut 2 pieces of stock 3/8" thick × 1/2" wide; saw 6-1/8" long.
4. Measure, mark, and cut 2 pieces of stock 3/8" thick × 3/4" wide; saw 6-1/8" long.
5. Make sure milling machine is in tram.
6. Deburr all over.
7. Place piece in mill vise on parallels with sawed end sticking out; tap down with dead blow hammer to seat part on parallels, then mill just enough off the end to clean it up.

8. Do this to all four pieces.

9. Deburr the end you just milled, flip part 180° on parallels in vise, tap down with dead blow hammer to seat part on parallels, and set your vise stop and mill to length.

10. Set the zero once the part is to size, then repeat with each part.

11. Place part in vise on parallels against vise stop; tap down with dead blow hammer to seat part on parallels, then set vise stop.

12. Paint part with layout dye and use height gage to lay out hole locations, being sure to leave grind stock on each end.

13. Insert edge finder in collet, set RPMs to 1,100, and edge find the X-axis and the Y-axis. Be sure to allow for 1/2 grind stock on the X-axis.

    a. X-axis should read −.1075 (this includes 1/2 of the grind stock).

    b. Y-axis should read .1075 (this includes 1/2 of the grind stock).

14. Place center drill in chuck and move to hole location.

15. Lower center drill to make sure it aligns with layout lines.

16. If location is correct, center drill hole at 1,200 RPMs, change to 1/8" drill, set the RPMs, and drill through part. Repeat process on second one of the same size.

17. Then change Y location for second set and center drill and drill each one.

18. Put a 1/4" endmill in collet and move to the location of the name pocket. Mill the name pocket .030 deep in each parallel.

MAKE SURE MAGNETIC CHUCK IS ON WHEN SURFACE GRINDING!

19. Surface grind all sides square and parallel. Deburr after each side is ground.

20. Surface grind the face to clean it up, deburr, then flip the part to grind bottom.

21. Deburr all edges, then place in grinder vise on parallels with ground sides against jaws.

22. Grind top of part to clean up; repeat for each parallel.

23. Remove from vise, deburr all edges, place ground side down on chuck with square blocking on each side, and grind side to clean it up. Grind them in pairs from this point on.

24. Deburr and place in large angle plate, secure against the inside corner of the angle plate with the compound clamps, and clean up end surface.

25. Turn end for end and repeat.

26. Now check hole locations and grind them into location and part to size. Be sure to deburr and block part when needed.

27. These must be flat and parallel. You might need to use shim stock to remove any bows in the parallels.

28. Grind a .015 × 45° chamfer on all edges.

29. Inspect the part to print dimensions; record on inspection sheet and turn in.

Send to Heat Treat

Student Name: _____ Date Submitted: _____

Class: _____ Total Hours on Job: _____

	Print Dimensions AND Tolerances	Student's Inspection	Instructor's Inspection	Instructor Comments
1	Are the parts chamfered and do they meet finish requirements?			
2	Are the parts properly identified?			
3	6" ± .015			
4	3/8" ± .015			
5	1/2" ± .015			
6	1/8" Thru Hole ± .015			
7	Location of 1/8" Hole .250 ± .005			
8	Name Slot .450 ± .005			
9	Name Slot .250 ± .005			
10	Name Slot .035 ± .005			
11	Both Parallels Ground to Match and Flat to ± .0002			
12	6" ± .015			
13	3/8" ± .015			
14	3/4" ± .015			
15	1/8" Thru Hole ± .015			
16	Location of 1/8" Hole .250 ± .005			
17	Name Slot .450 ± .005			
18	Name Slot .250 ± .005			
19	Name Slot .035 ± .005			
20	Both Parallels Ground to Match and Flat to ± .0002			

PROJECT GRADE: _____

Mill Angles

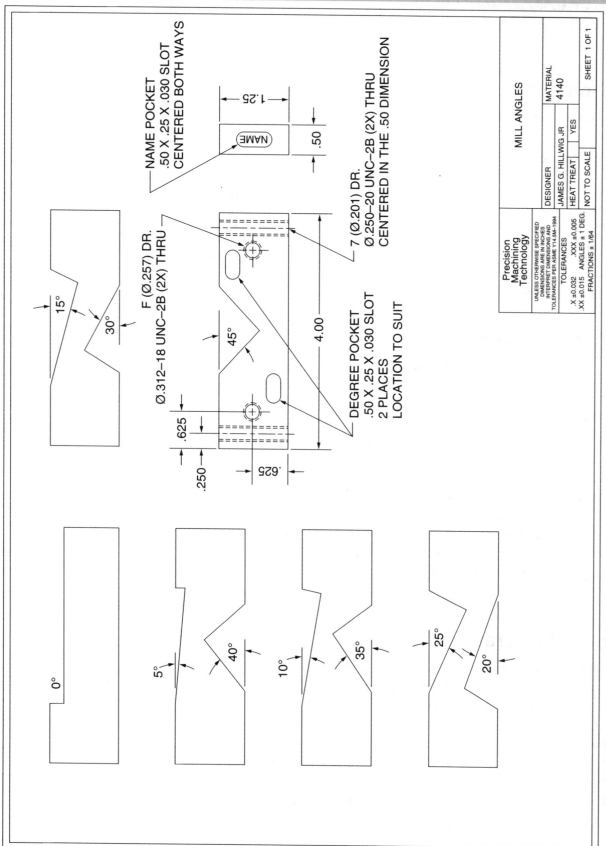

NAME POCKET
.50 X .25 X .030 SLOT
CENTERED BOTH WAYS

NAME

1.25

.50

7 (Ø.201) DR.
Ø.250–20 UNC–2B (2X) THRU
CENTERED IN THE .50 DIMENSION

F (Ø.257) DR.
Ø.312–18 UNC–2B (2X) THRU

DEGREE POCKET
.50 X .25 X .030 SLOT
2 PLACES
LOCATION TO SUIT

15°

30°

45°

4.00

.625

.625

.250

0°

5°

40°

10°

35°

25°

20°

Precision Machining Technology		MILL ANGLES		
UNLESS OTHERWISE SPECIFIED DIMENSIONS ARE IN INCHES INTERPRET DIMENSIONS AND TOLERANCES PER ASME Y14.5M–1994		DESIGNER		MATERIAL
		JAMES G. HILLWIG JR		4140
TOLERANCES		HEAT TREAT	YES	
.X ±0.032 .XXX ±0.005				
.XX ±0.015 ANGLES ± 1 DEG.		NOT TO SCALE		SHEET 1 OF 1
FRACTIONS ± 1/64				

NIMS Skills Practiced

- Job Process Planning
- Layout
- Surface Grinding and Safety
- Vertical Milling and Safety
- Squaring up a Block

Equipment

Power saw, milling machine, surface grinder

Tools Needed

Scribe	6" steel rule	Layout dye
Dead blow hammer	File	File card
#7 letter "F" drill bits	Various sizes of endmills	Parallels
Edge finder	Center drill	Vise stop
Grinding wheel	Grinder vise	Stamps
Tramming items for mill	1/4 -20 tap	.312-18 tap
Adjustable mill angle	Magnetic sine chuck	

Material Needed

- 7 @ .5" × 1.25" × 4-1/8" tool steel (oversize stock)

Safety

1. Follow all shop safety rules given to you by your instructor and be sure to pass any safety tests given before using shop equipment.
2. Follow all safety rules for power equipment.
3. Be careful of sharp tools when using them.
4. Make sure the file has a handle on it.
5. Be sure all cutting tools are sharp and care is taken while handling them.
6. Be sure ways are locked on mill when necessary.
7. Ring grinding wheels before using them.
8. Make sure magnetic chuck is on when surface grinding.

Order of Operations

1. Look over blueprint to determine tolerances, tools, and material needed.
2. Gather tools needed, figure out RPMs for required tools, and write down their RPMs on sheet.
3. Measure, mark, and cut 7 pieces of stock.
4. Make sure milling machine is in tram.
5. Deburr all over.
6. Place piece in mill vise on parallels, square it up, and take it to size.
7. Do this to all 7 pieces.
8. Lay out hole locations, being sure to leave grind stock on each end.
9. Place part in vise and edge find.

10. Place center drill in chuck and center drill holes.

11. Swap blocks and repeat process until all holes are drilled.

12. Now chamfer each hole.

13. Now tap each hole to correct size.

14. Lay out the angles on each part and write the angle for each angle so you don't get confused.

15. Place each part in the vise on the adjustable mill angle and mill each angle to print dimensions.

16. Cut the angle pockets and name pockets in each piece where indicated and stamp them accordingly.

MAKE SURE MAGNETIC CHUCK IS ON WHEN SURFACE GRINDING!

1. Surface grind all sides square and parallel. Deburr after each side is ground.

2. These must be flat, square, and parallel.

3. Using a cup wheel or dished-out wheel, place parts on magnetic sine chuck and grind each of the angles.

4. Grind a .015 × 45° chamfer on all outside edges. Do not chamfer the angles.

5. Inspect the part to print dimensions; record on inspection sheet and turn in.

Send to Heat Treat

Student Name: _____ Date Submitted: _____

Class: _____ Total Hours on Job: _____

	Print Dimensions AND Tolerances	Student's Inspection	Instructor's Inspection	Instructor Comments
1	Is the part free of burrs and does it meet finish requirements?			
2	Is the part properly identified?			
3	**Length** 4. ± .015			
4	**Thickness** .50 ± .015			
5	**Height** 1.25 ± .015			
6	**.312-18 Holes** .625 × .625 ± .005			
7	**1/4-20 Holes** .250 × .250 ± .005			
8	**Angle** 0° ± 1/2°			
9	**Angle** 5° ± 1/2°			
10	**Angle** 40° ± 1/2°			
11	**Angle** 10° ± 1/2°			
12	**Angle** 35° ± 1/2°			
13	**Angle** 25° ± 1/2°			
14	**Angle** 20° ± 1/2°			
15	**Angle** 15° ± 1/2°			
16	**Angle** 30° ± 1/2°			
17	**Angle** 45° ± 1/2°			
18	**Angle** 45° ± 1/2°			

PROJECT GRADE: _____

1-2-3 Block

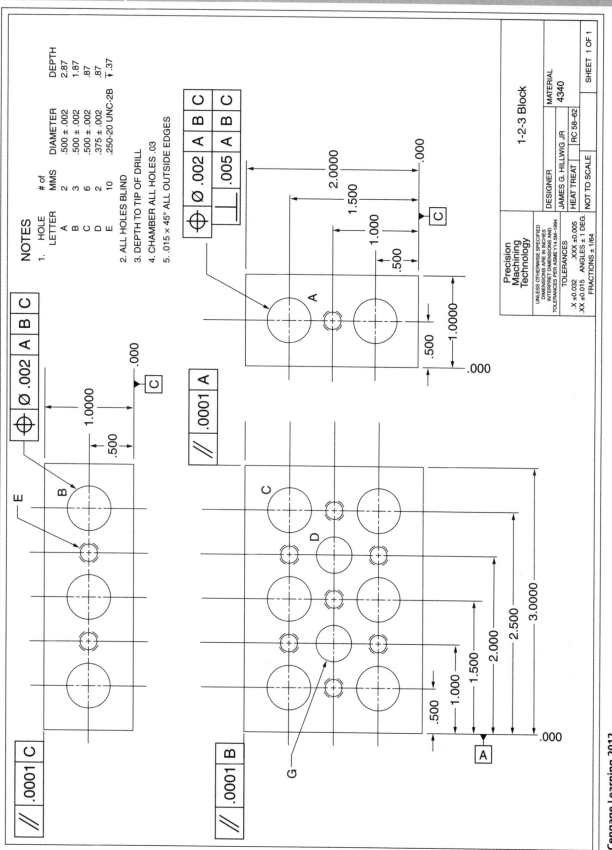

NOTES

1.

HOLE LETTER	# of MMS	DIAMETER	DEPTH
A	2	.500 ± .002	2.87
B	3	.500 ± .002	1.87
C	6	.500 ± .002	.87
D	2	.375 ± .002	.87
E	10	.250-20 UNC-2B	▼ .37

2. ALL HOLES BLIND

3. DEPTH TO TIP OF DRILL

4. CHAMBER ALL HOLES .03

5. .015 × 45° ALL OUTSIDE EDGES

Precision Machining Technology			1-2-3 Block	
UNLESS OTHERWISE SPECIFIED DIMENSIONS ARE IN INCHES INTERPRET DIMENSIONS PER ASME Y14.5M–1994		DESIGNER	MATERIAL	
		JAMES G. HILLWIG JR	4340	
TOLERANCES		HEAT TREAT	RC 58–62	
.X ±0.032 .XXX ±0.005				
.XX ±0.015 ANGLES ± 1 DEG.		NOT TO SCALE		
FRACTIONS ± 1/64			SHEET 1 OF 1	

NIMS Skills Practiced

- Job Process Planning
- Layout
- Surface Grinding and Safety
- Vertical Milling and Safety
- Squaring up a Block

Equipment

Power saw, milling machine, surface grinder

Tools Needed

Scribe
Dead blow hammer
Various sizes of drill bits
Edge finder
Grinding wheel
Tramming items for mill

6" steel rule
File
Various sizes of endmills
Center drill
Grinder vise

Layout dye
File card
Parallels
Vise stop
Stamps

Material Needed

- 2 @ 1" × 2" × 3-1/8" tool steel (oversize stock)

Safety

1. Follow all shop safety rules given to you by your instructor and be sure to pass any safety tests given before using shop equipment.
2. Follow all safety rules for power equipment.
3. Be careful of sharp tools when using them.
4. Make sure the file has a handle on it.
5. Be sure all cutting tools are sharp and care is taken while handling them.
6. Be sure ways are locked on mill when necessary.
7. Ring grinding wheels before using them.
8. Make sure magnetic chuck is on when surface grinding.

Order of Operations

1. Look over blueprint to determine tolerances, tools, and material needed.
2. Gather tools needed, figure out RPMs for required tools, and write down their RPMs on sheet.
3. Measure, mark, and cut 2 pieces of stock 1" thick × 2" wide; saw 3-1/8" long.
4. Make sure milling machine is in tram.
5. Deburr all over.
6. Place piece in mill vise on parallels with sawed end sticking out; tap down with dead blow hammer to seat part on parallels, then mill just enough off the end to clean it up.
7. Do this to both pieces.
8. Deburr the end you just milled, flip part 180° on parallels in vise, tap down with dead blow hammer to seat part on parallels, set your vise stop, and mill to length.

9. Set the zero once the part is to size, then mill next part to same zero.

10. Paint part with layout dye and use height gage to lay out hole locations, being sure to leave grind stock on each end.

11. Place part in vise on parallels against vise stop, tap down with dead blow hammer to seat part vise bottom so you can drill side #1, then set vise stop.

12. Insert edge finder in collet, set RPMs to 1,100, and edge find the X-axis and the Y-axis. Be sure to allow for 1/2 grind stock on the X- and Y-axis.

    a. X-axis should read $-.1075$ (this includes 1/2 of the grind stock).

    b. Y-axis should read .1075 (this includes 1/2 of the grind stock).

13. Place center drill in chuck and move to first letter "D" hole location.

14. Lower center drill to make sure it aligns with layout lines.

15. If location is correct, center drill hole at 1,200 RPMs.

16. Move to center hole and center drill.

17. Move to third hole and center drill.

18. Swap blocks and repeat process.

19. Now replace center drill with #7 drill and drill each hole again to depths required on print.

20. Now, using a 3/8" drill, drill letter "D" holes to depth in both blocks.

21. Replace with a 1/2" drill, then drill letter "D" holes to depth in both blocks.

22. Now repeat this process on side #2 of both blocks, then side #3, being careful not to break through the bottoms.

23. If you take a 1/2" endmill you can flatten out the bottoms of the three letter "F" holes and stamp your initials inside of the blocks before heat treatment.

24. Now chamfer each hole.

25. Now tap each 1/4-20 hole.

MAKE SURE MAGNETIC CHUCK IS ON WHEN SURFACE GRINDING!

26. Surface grind all sides square and parallel. Deburr after each side is ground.

27. Surface grind face #3 of both blocks together to clean them up, deburr, then flip the parts to grind bottom.

28. Deburr all edges, then place one at a time in grinder vise with ground sides against jaws, leaving side #1 sticking enough out of side of vise that you can place vise on its side to grind end. Grind side #1.

29. Place vise on bottom and grind top of part to clean up side #2.

30. Repeat with second block.

31. Remove from vise, deburr all edges, place ground sides of both blocks down on chuck with square blocking on each side, and grind side to clean it up. Grind them in pairs from this point on.

32. Deburr and place last side up with square blocking on both sides, then clean up end surface.

33. Now check hole locations and grind them into location and part to size. Be sure to deburr and block part when needed.

34. These must be flat, square, and parallel.

35. Grind a .015 $\times$ 45° chamfer on all edges.

36. Inspect the part to print dimensions; record on inspection sheet and turn in.

Send to Heat Treat

Student Name: _____ Date Submitted: _____

Class: _____ Total Hours on Job: _____

		Print Dimensions AND Tolerances	Student's Inspection	Instructor's Inspection	Instructor Comments
1		Is the part free of burrs and does it meet finish requirements?			
2		Is the part properly identified?			
3		{Length} 3.000 +.0005 −.0000			
4		{Width} 2.000 +.0005 −.0000			
5		{Height} 1.000 +.0005 −.0000			
6		{D Holes} 1/2 Diameter ± .015			
7		{D Holes} 2 7/8 Deep ± .015			
8		{E Holes} 1/2 Diameter ± .015			
9		{E Holes} 1 7/8 Deep ± .015			
10		{F Holes} 1/2 Diameter ± .015			
11		{F Holes} 7/8 Deep ± .015			
12		{G Holes} 3/8 Diameter ± .015			
13		{G Holes} 7/8 Deep ± .015			
14		{H Holes} 1/4-20 Tap			
15		{H Holes} 3/8 Deep ± .015			
16		{Chamfer Outside} .015 ± .015			
17		{Holes Chamfered 1/32 ± .015}			
18		Hole locations in center?			

PROJECT GRADE: _____

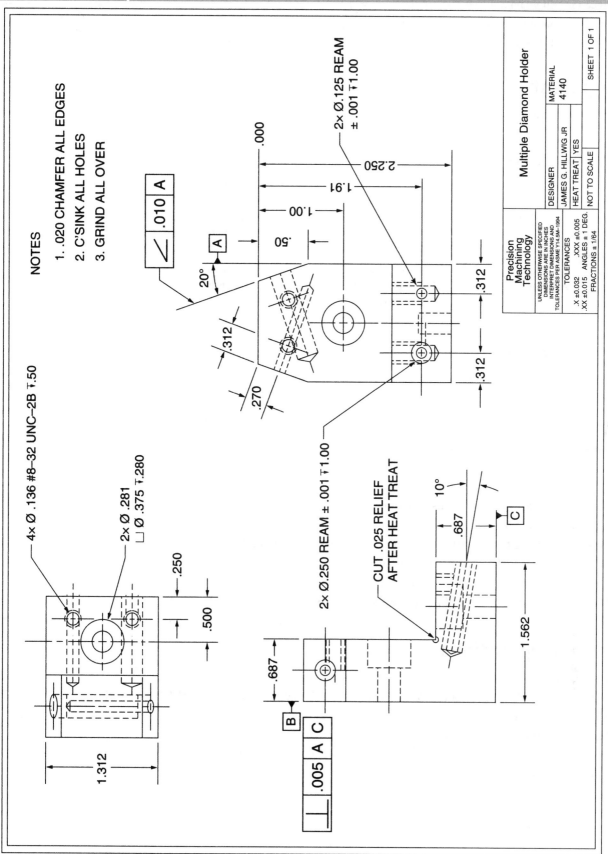

NOTES

1. .020 CHAMFER ALL EDGES
2. C'SINK ALL HOLES
3. GRIND ALL OVER

2× Ø.125 REAM ±.001 ⊤1.00

2× Ø.250 REAM ±.001 ⊤1.00

CUT .025 RELIEF AFTER HEAT TREAT

4× Ø .136 #8–32 UNC–2B ⊤.50

2× Ø .281
⌴ Ø .375 ⊤.280

Multiple Diamond Holder

Precision Machining Technology	DESIGNER	MATERIAL
UNLESS OTHERWISE SPECIFIED DIMENSIONS ARE IN INCHES INTERPRET DIMENSIONS AND TOLERANCES PER ASME Y14.5M–1994	JAMES G. HILLWIG JR	4140
TOLERANCES	HEAT TREAT	YES
.X ±0.032 XXX ±0.005 .XX ±0.015 ANGLES ± 1 DEG. FRACTIONS ± 1/64	NOT TO SCALE	SHEET 1 OF 1

NIMS Skills Practiced

- Job Process Planning
- Layout
- Surface Grinding and Safety
- Vertical Milling and Safety
- Squaring up a Block

Equipment

Power saw, milling machine, surface grinder

Tools Needed

Scribe	6" steel rule	Layout dye
Dead blow hammer	File	File card
Various sizes of drill bits	Various sizes of endmills	Stamps
Various sizes of reamers	Parallels	Vise stop
Edge finder	Center drill	Cup wheel
Adjustable mill angle	Grinding wheel	Grinder vise
Tramming items for mill	Magnetic sine plate	Counterbore
8-32 tap	Countersink	

Material Needed

- 1.5" × 1.75" × 2.375" tool steel

Safety

1. Follow all shop safety rules given to you by your instructor and be sure to pass any safety tests given before using shop equipment.
2. Follow all safety rules for power equipment.
3. Be careful of sharp tools when using them.
4. Make sure the file has a handle on it.
5. Be sure all cutting tools are sharp and care is taken while handling them.
6. Be sure ways are locked on mill when necessary.
7. Ring grinding wheels before using them.
8. Make sure magnetic chuck is on when surface grinding.

Order of Operations

1. Look over blueprint to determine tolerances, tools, and material needed.
2. Gather tools needed, figure out RPMs for required tools, and write down their RPMs on sheet.
3. Measure, mark, and cut stock 1.5" thick × 1.75" wide; saw 2.375" long.
4. Make sure milling machine is in tram.
5. Deburr all over.
6. Mill part square.
7. Lay out "L" shape of part.

8. Lay the part on its back in the vise on parallels and mill out the "L" shape, leaving .015" total for grind stock on each .687 dimension.

9. Lay out the rest of the features you need to complete the part.

10. Deburr and place side of part on adjustable mill angle.

11. Mill the 20° angles on part, edge find, then drill the holes in the following order.

12. BE CAREFUL TO DRILL AND REAM THE .125 HOLE FIRST IN THE PART!

13. Drill and ream the .250 hole next.

14. Now repeat the same process for the holes on the 10° surface.

15. Drill, countersink, and carefully tap the 8-32 holes.

16. Drill and counterbore the two holes as shown.

17. Mill a name pocket on the part and stamp your initials in it.

MAKE SURE MAGNETIC CHUCK IS ON WHEN SURFACE GRINDING!

18. Sandblast part to clean it up for grinding.

19. Using the cutoff wheel, grind the relief in the corner as shown.

20. Surface grind all sides square and parallel; deburr after each side is ground.

21. Grind angles using the magnetic sine bar.

22. With your instructor's assistance, place a cup wheel on the surface grinder and grind the inside of the "L."

23. Now check hole locations and grind them into location and part to size. Be sure to deburr and block part when needed.

24. Grind a .015 × 45° chamfer on all edges.

25. You will need the instructor's help to set up the compound angle needed to grind the inside chamfers.

26. Inspect the part to print dimensions; record on inspection sheet and turn in.

Send to Heat Treat

Student Name: _____ Date Submitted: _____

Class: _____ Total Hours on Job: _____

	Print Dimensions AND Tolerances	Student's Inspection	Instructor's Inspection	Instructor Comments
1	Is the part free of burrs and does it meet finish requirements?			
2	Is the part properly identified?			
3	**Front View** 1.562 ± .005 length			
4	.687 ± .005 length (top)			
5	.687 ± .005 length (right side)			
6	10° angle (2 holes)			
7	**Right Side View** 2.250 ± .005 length			
8	length ±.015			
9	.50 length ±.015			
10	.34 length ±.015			
11	.312 ± .005 length (left side)			
12	.312 ± .005 length (right side)			
13	20° angle (2 places)			
14	# 8-32 THD. .500 DP .360 × .382 loc.			
15	# 8-32 THD. .500 DP .360 × .930 loc.			
16	Ø.125 ± .005 1.00 DP (2 holes)			
17	Ø.250 ± .005 1.00 DP (2 holes)			
18	Counterbore for .250 Allen Screw 1.00 × .656 loc.			
19	**Top View** 1.312 ± .005 length			
20	Ø.250 ± .005 1.00 DP			
21	Ø.125 ± .005 1.00 DP			
22	Counterbore for .250 Allen Screw .500 × .656 loc.			
23	# 8-32 THD. .500 DP .250 ×.312 loc. (top hole)			
24	# 8-32 THD. .500 DP .250 × .312 loc. (bottom hole)			
25	.020 Chamfer All Edges			

PROJECT GRADE: _____

Male/Female Tap Center

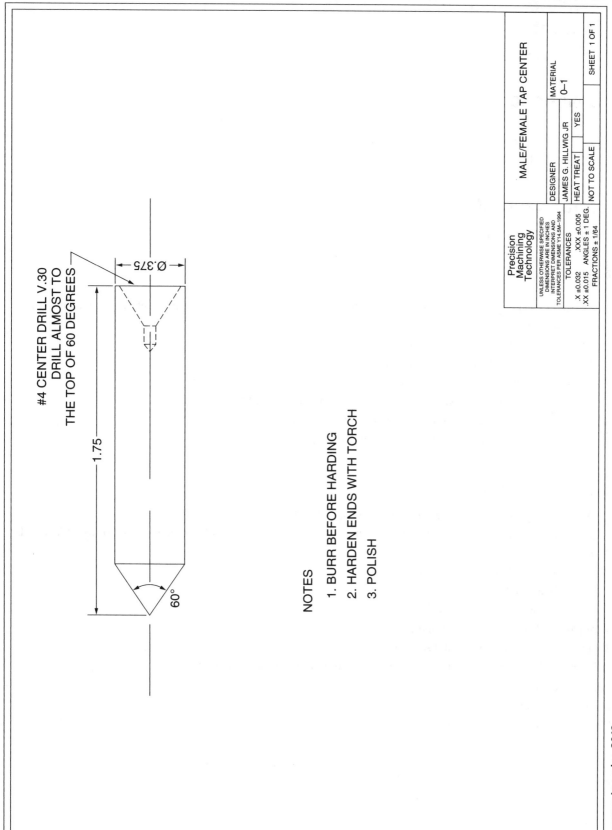

#4 CENTER DRILL V.30
DRILL ALMOST TO
THE TOP OF 60 DEGREES

Ø.375

1.75

60°

NOTES

1. BURR BEFORE HARDING

2. HARDEN ENDS WITH TORCH

3. POLISH

Precision Machining Technology	MALE/FEMALE TAP CENTER		
UNLESS OTHERWISE SPECIFIED DIMENSIONS ARE IN INCHES INTERPRET DIMENSIONS AND TOLERANCES PER ASME Y14.5M–1994	DESIGNER	MATERIAL	
	JAMES G. HILLWIG JR	0–1	
TOLERANCES	HEAT TREAT	YES	
.X ±0.032 .XXX ±0.005			
.XX ±0.015 ANGLES ± 1 DEG.	NOT TO SCALE		SHEET 1 OF 1
FRACTIONS ± 1/64			

NIMS Skills Practiced

• Job Process Planning
• Turning Operations: Chucking

Equipment

Power saw, lathe, bench grinder

Tools Needed

Scribe
File card
Turning, facing tools

6" steel rule
Various files

Layout dye
Center drill

Material Needed

• 0-1 tool steel 3/8" × 1-5/8"

Safety

1. Follow all shop safety rules given to you by your instructor and be sure to pass any safety tests given before using shop equipment.
2. Follow all safety rules for power equipment.
3. Be careful of sharp tools when using them.
4. Make sure the files have a handle on them.
5. Be sure all cutting tools are sharp and care is taken while handling them.
6. Be sure chuck is tightened and chuck wrench is never left in chuck.
7. Make sure stock is supported with center when necessary.

Order of Operations

1. Look over blueprint to determine tolerances, tools, and material needed.
2. Gather tools needed, figure out RPMs for required stock diameters and drills used, and write down their RPMs on sheet.
3. Measure and cut stock sizes needed.
4. Place stock in chuck.
5. Face off one end and center drill.
6. Turn part in chuck and face to length.
7. Set the compound rest to 30° and set tool.
8. Using a steady feed with the compound rest, turn until angle comes to a point.
9. Polish taper and rest of part.
10. Inspect part and turn in.
11. Heat treat part with torches and re-polish.

Student Name: _____ Date Submitted: _____

Class: _____ Total Hours on Job: _____

	Print Dimensions AND Tolerances	Student's Inspection	Instructor's Inspection	Instructor Comments
1	Is the part free of burrs?			
2	Is the part properly identified?			
3	3/8" diameter			
4	1-3/4" long			
5	Is the #4 center-drilled hole good?			
6	Is the 60-degree angle good?			
7	Does the part meet finish requirements?			
8				
9				
10				
11				
12				
13				
14				
15				

PROJECT GRADE: _____

Lathe Chuck Center

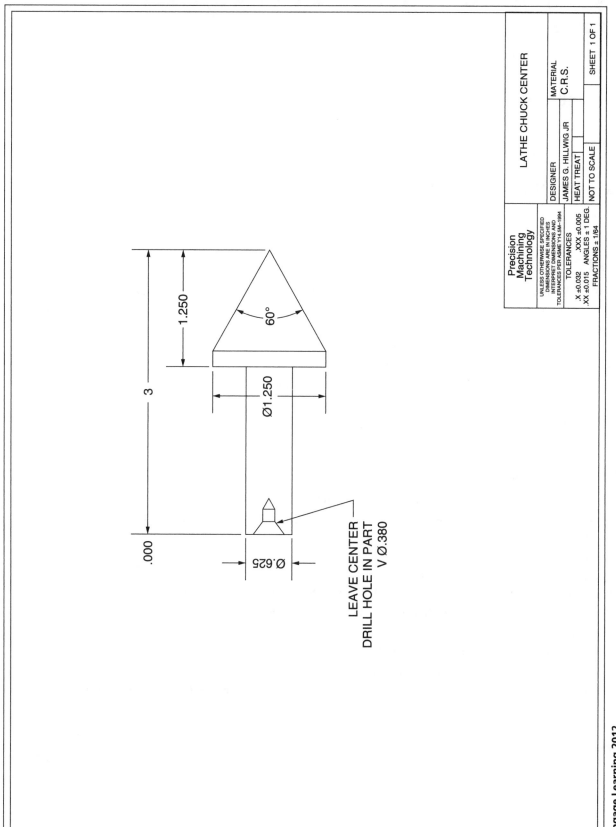

LEAVE CENTER
DRILL HOLE IN PART
V Ø.380

Ø.625

.000

3

Ø1.250

1.250

60°

Precision Machining Technology

UNLESS OTHERWISE SPECIFIED
DIMENSIONS ARE IN INCHES
INTERPRET DIMENSIONS AND
TOLERANCES PER ASME Y14.5M–1994

TOLERANCES
.X ±0.032   .XXX ±0.005
.XX ±0.015   ANGLES ± 1 DEG.
FRACTIONS ± 1/64

DESIGNER
JAMES G. HILLWIG JR

HEAT TREAT

NOT TO SCALE

MATERIAL
C.R.S.

LATHE CHUCK CENTER

SHEET 1 OF 1

NIMS Skills Practiced

- Job Process Planning
- Turning Operations: Chucking

Equipment

Power saw, lathe, bench grinder

Tools Needed

6" steel rule   File card   Various files
Center drill   Various types of turning and facing tools

Material Needed

- 1-1/4" diameter × 3-1/8" long CRS

Safety

1. Follow all shop safety rules given to you by your instructor and be sure to pass any safety tests given before using shop equipment.
2. Follow all safety rules for power equipment.
3. Be careful of sharp tools when using them.
4. Make sure the files have a handle on them.
5. Be sure all cutting tools are sharp and care is taken while handling them.
6. Be sure chuck is tightened and chuck wrench is never left in chuck.
7. Make sure stock is supported with center when necessary.

Order of Operations

1. Look over blueprint to determine tolerances, tools, and material needed.
2. Gather tools needed, figure out RPMs for required stock diameters and drills used, and write down their RPMs on sheet.
3. Measure and cut stock sizes needed.
4. Place in chuck and face and center drill one end.
5. Slide part 2" out of chuck and place tailstock's live center in center-drilled hole.
6. Turn the .501 diameter, leaving a few thousands of an inch to polish to size.
7. Polish to size and finish.
8. Turn part end for end and machine to length.
9. Set up compound rest and turn the 60° included angle.
10. Do not remove or loosen the part from the chuck until angle is complete; if you do, you will need to re-cut the 60° angle.
11. Polish the angle.
12. Fill out inspection sheet and turn in.

Student Name: _____ Date Submitted: _____

Class: _____ Total Hours on Job: _____

	Print Dimensions AND Tolerances	Student's Inspection	Instructor's Inspection	Instructor Comments
1	Is the part free of burrs and does it meet finish requirements?			
2	Is the part properly identified?			
3	3" ± .015			
4	1-1/4" ± .015			
5	1-1/4" ± .015			
6	.625 ± .005			
7	60° ± 1°			
8	1-3/4" ± .015			
9	1-1/4" dimension polished			
10	.625 dimension polished			
11				
12				
13				
14				
15				

PROJECT GRADE: _____

Punch Set

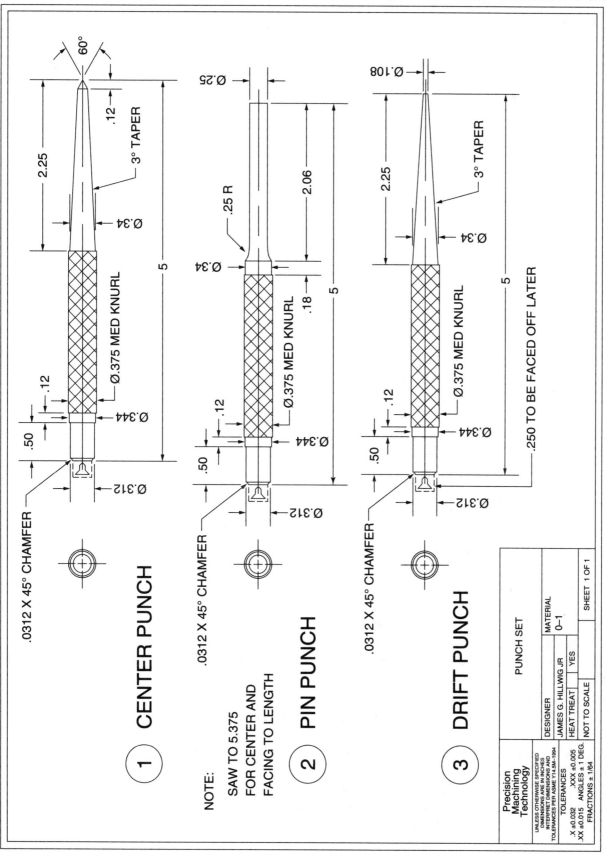

① CENTER PUNCH

60°

2.25 .12

3° TAPER

Ø.34

5

Ø.375 MED KNURL

.12

Ø.344

.50

Ø.312

.0312 X 45° CHAMFER

NOTE:

SAW TO 5.375
FOR CENTER AND
FACING TO LENGTH

② PIN PUNCH

Ø.25

.25 R

Ø.34

2.06

5

Ø.375 MED KNURL

.18

.12

Ø.344

.50

Ø.312

.0312 X 45° CHAMFER

③ DRIFT PUNCH

Ø.108

2.25

3° TAPER

Ø.34

5

Ø.375 MED KNURL

.12

Ø.344

.50

Ø.312

.250 TO BE FACED OFF LATER

.0312 X 45° CHAMFER

Precision Machining Technology		PUNCH SET	
UNLESS OTHERWISE SPECIFIED DIMENSIONS ARE IN INCHES INTERPRET DIMENSIONS AND TOLERANCES PER ASME Y14.5M–1994	DESIGNER	MATERIAL	
TOLERANCES .X ±0.032 .XXX ±0.005 .XX ±0.015 ANGLES ± 1 DEG. FRACTIONS ± 1/64	JAMES G. HILLWIG JR	0–1	
	HEAT TREAT	YES	
	NOT TO SCALE		SHEET 1 OF 1

NIMS Skills Practiced

- Job Process Planning
- Turning Operations: Chucking

Equipment

Power saw, lathe, bench grinder

Tools Needed

Marker
Various files
Various grits of sandpaper
Radius gages

6" steel rule
Center drill
Turning and facing tools

File card
Knurling tool
Radius tool

Material Needed

- 3/8" diameter × 5-3/8" long

Safety

1. Follow all shop safety rules given to you by your instructor and be sure to pass any safety tests given before using shop equipment.
2. Follow all safety rules for power equipment.
3. Be careful of sharp tools when using them.
4. Make sure the files have a handle on them.
5. Be sure all cutting tools are sharp and care is taken while handling them.
6. Be sure chuck is tightened and chuck wrench is never left in chuck.
7. Make sure stock is supported with center when necessary.

Order of Operations

1. Look over blueprint to determine tolerances, tools, and material needed.
2. Gather tools needed, figure out RPMs for required stock diameters and drills used, and write down their RPMs on sheet.
3. Measure and cut stock sizes needed.
4. Place in chuck and face and center drill one end no deeper than 1/4" deep. This will be used for knurling only and removed afterward. Do this on each part.
5. Now slide part out of chuck, leaving approx. 1-1/2" in chuck, placing the tailstock in the center-drilled hole.
6. Using the medium knurl found on the knurling tool, knurl the .375 diameter from the chuck toward the tailstock. Make sure everything is tight.
7. Repeat on the other two parts.
8. Face out the center-drilled hole of each part just enough to remove it, then flip parts end for end and take them to size, 5".
9. Place parts, one at a time, back in chuck with knurled side sticking out 1"; protect the knurls from being crushed in the chuck.
10. Turn the 11/32 diameter back 5/8" on each part, leaving a few thousandths to polish.
11. Now turn the .312 diameter 1/2" on each part, leaving a few thousandths to polish.
12. Using a file, put the chamfer on each part.

13. Flip parts end for end and turn the 11/32 diameter back on each punch 2-1/4", leaving a few thousandths for polishing.

14. Set the taper on the compound rest at 3° and turn the taper on the center punch and the drift punch; leave a few thousandths for polishing.

15. Using a file, file the 60° angle on the end of the center punch.

16. Place the pin punch in the chuck and turn the .250 diameter back 2-1/16" minus 1/4" (to leave stock for the radius). Leave a few thousandths on the .250 diameter for polishing.

17. Place the 1/4" radius tool in holder and blend the radius in.

18. Polish all the turned surfaces to remove any marks or scratches before heat treatment or they will not come out easily.

Send to Heat Treat

19. Polish all the turned surfaces, leaving the black in the knurls, then dip them in oil and the black will contrast nicely.

20. Wipe off excess oil; fill out inspection sheet and turn in.

Student Name: _____ Date Submitted: _____

Class: _____ Total Hours on Job: _____

	Print Dimensions AND Tolerances	Student's Inspection	Instructor's Inspection	Instructor Comments
1	Is the part free of burrs and does it meet finish requirements?			
2	Is the part properly identified?			
3	**{Center Punch}** 5." ± .015			
4	(Diameter) .312 ± .005			
5	(Length of .312) 1/2 ± .015			
6	(Diameter) 11/32 ± .015			
7	(Length of 11/32) 1/8 ± .015			
8	(Diameter/tapered end) 11/32 ± .015			
9	(Length of 11/32 tapered end) 2-1/4 ± .015			
10	3° Taper ± 1°			
11	60° Taper ± 1°			
12	45° Chamfer by 1/32 ± .015			
13	(Taper) 1/8 ± .015			
14	**{Pin Punch}** 5." ± .015			
15	(Diameter) .312 ± .005			
16	(Length of .312) 1/2 ± .015			
17	(Diameter) 11/32 ± .015			
18	(Length of 11/32) 1/8 ± .015			
19	(Diameter/.250 end) 11/32 ± .015			
20	(Length of 11/32 .250 end) 3/16 ± .015			
21	(Length of .250 end) 2-1/16 ± .015			

	Print Dimensions AND Tolerances	Student's Inspection	Instructor's Inspection	Instructor Comments
22	(Diameter of .250 end) .250 ± .005			
23	1/4 Radius ± .015			
24	45° Chamfer by 1/32 ± .015			
25	**{Drift Punch}** 5." ± .015			
26	(Diameter) .312 ± .005			
27	(Length of .312) 1/2 ± .015			
28	(Diameter) 11/32 ± .015			
29	(Length of 11/32) 1/8 ± .015			
30	(Diameter/tapered end) 11/32 ± .015			
31	(Length of 11/32 tapered end) 2-1/4 ± .015			
32	3° Taper ± 1°			
33	(Tapered end) .108 Diameter ± .005			
34	45° Chamfer by 1/32 ± .015			

PROJECT GRADE: _____

Reversible Pocket Scribe

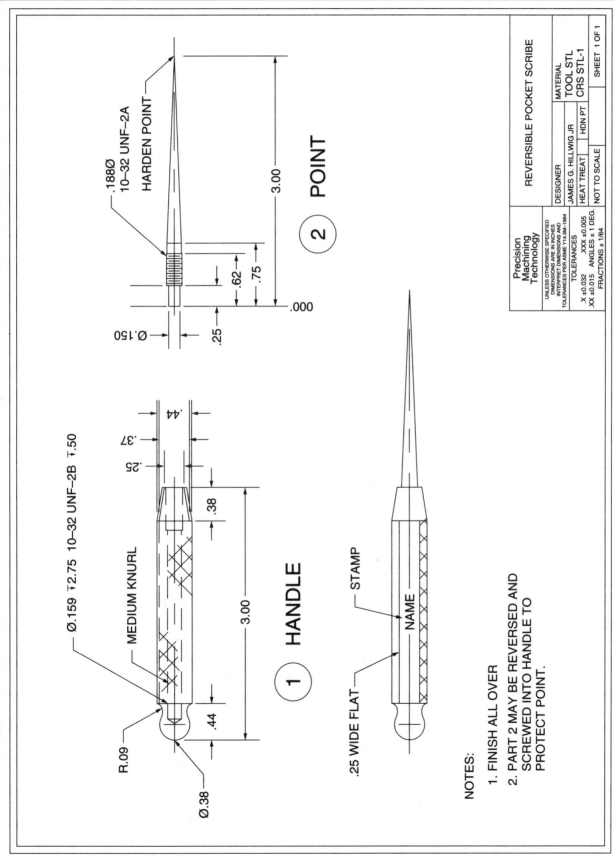

.188Ø
10–32 UNF–2A
HARDEN POINT

Ø.150

.25

.62

.75

3.00

.000

② POINT

Ø.159 ⌴2.75 10–32 UNF–2B ⌴.50

.44

.37

.25

.38

MEDIUM KNURL

3.00

R.09

Ø.38

.44

① HANDLE

STAMP

NAME

.25 WIDE FLAT

NOTES:

1. FINISH ALL OVER

2. PART 2 MAY BE REVERSED AND SCREWED INTO HANDLE TO PROTECT POINT.

Precision Machining Technology		DESIGNER	MATERIAL
		JAMES G. HILLWIG JR	TOOL STL
UNLESS OTHERWISE SPECIFIED DIMENSIONS ARE IN INCHES INTERPRET DIMENSIONS PER ASME Y14.5M–1994	REVERSIBLE POCKET SCRIBE	HEAT TREAT HDN PT	CRS STL-1
TOLERANCES .X ±0.032 .XXX ±0.005 .XX ±0.015 ANGLES ± 1 DEG. FRACTIONS ± 1/64		NOT TO SCALE	SHEET 1 OF 1

NIMS Skills Practiced

- Job Process Planning
- Turning Operations: Chucking

Equipment

Power saw, lathe, bench grinder

Tools Needed

Marker	6" steel rule	File card
Various files	Center drill	Knurling tool
#21 drill (short and long)	10-31 tap and die	Various grits of sandpaper
Various types of turning and facing tools	Radius gages	

Material Needed

- 1/2" diameter × 4" long CRS or aluminum (handle)
- 3/16" × 3" long tool steel (Point: this needs to be heat treated 0-1; can be done with torches)

Safety

1. Follow all shop safety rules given to you by your instructor and be sure to pass any safety tests given before using shop equipment.
2. Follow all safety rules for power equipment.
3. Be careful of sharp tools when using them.
4. Make sure the files have a handle on them.
5. Be sure all cutting tools are sharp and care is taken while handling them.
6. Be sure chuck is tightened and chuck wrench is never left in chuck.
7. Make sure stock is supported with center when necessary.

Order of Operations

HANDLE

1. Look over blueprint to determine tolerances, tools, and material needed.
2. Gather tools needed, figure out RPMs for required stock diameters and drills used, and write down their RPMs on sheet.
3. Measure and cut stock sizes needed.
4. Place in chuck and face and center drill one end; make sure the center-drilled hole is no larger than 3/16" in diameter. This will be the end that the taper goes on and the #21 drilled hole is in.
5. Turn part end for end (stick center-drilled end into chuck); face and turn to a .400 diameter by .375 long.
6. This step is only done to create a way to hold the part in the chuck so you can safely knurl the part. (It will be cut off later.)
7. Place the .400 diameter you just turned into the chuck. Place the lathe center in the tailstock; place the point of the lathe center into the center-drilled hole you made earlier, which was no larger than 3/16". Now turn the part to .437 diameter back 3-1/2" long. This will become the knurled surface.
8. Using the medium knurl found on the knurling tool, knurl the .437 diameter you just turned as far as you safely can without hitting the chuck. Remember, looking at the print, you'll see one end of the handle has a 3/8" long taper and the other end has a 3/8" long ball, so don't worry if the knurl doesn't reach both edges of the part.

9. Leaving the part setup as is, remove the knurling tool and put your cutting tool back in. Turn down the end with the center-drilled hole in it, leaving the lathe center in place, to a .375 diameter × .375 long. This is the end the taper will go on.

10. Using simple trig, trig out the angle you need to set the compound rest. See instructor for formula and proper setup of compound rest. Using the handle on the compound rest, turn the taper, leaving a few thousandths for filing and polishing.

11. Turn part end for end, holding the knurls in the jaws (place something on knurls to protect them; see your instructor); face part to 3" long. You will be removing the .400 diameter × .375 end used to hold the part for knurling.

12. Once the part is 3" long, turn .375 diameter back 7/16" long on the same end; this will become the ball on the handle.

13. Use various files to form the ball. File the ball on the part using radius gages to check it. Leave a little for polishing.

14. Turn part end for end, placing ball end in jaws and holding part by knurls (use same protector on the knurls); using the drill chuck in the tailstock, drill with a #21 drill 3/4 of the way up the flutes with a regular-length drill bit. Then see instructor for the extra-long #21 drill bit to drill hole to depth.

15. Tap the hole to print; ask instructor for help to do it on the machine.

16. Polish the taper and ball.

17. You can grind the flat on the handle (using the surface grinder and vise with parallels) according to print, and then stamp your name on the flat if required by instructor.

POINT

1. Use 3/16" tool steel.

2. File .150 diameter by .250 long on one end.

3. Turn end for end in chuck, leaving about 1" sticking out to file to the tapered point as shown on print.

4. Keep sliding out a little at a time until .375 of the original 3/16" remains (this is for the threads).

5. Compare your piece to the print—it should look identical.

6. Now remove the point and hold in vise with soft jaws and thread according to print.

7. Polish.

8. Heat treat.

9. Polish.

Student Name: _____ Date Submitted: _____

Class: _____ Total Hours on Job: _____

	Print Dimensions AND Tolerances	Student's Inspection	Instructor's Inspection	Instructor Comments
1	Is the part free of burrs and does it meet finish requirements?			
2	Is the part properly identified?			
3	**{Handle}** 3.00 ± .015			
4	{Ball} .38 Diameter ± .015			
5	{Ball} .09 Radius ± .015			
6	{Ball} .44 ± .015			
7	{Knurl} .44 ± .015			
8	{Taper} .375 ± .005			
9	{Taper} .250 ± .005			
10	{Taper} .38 ± .015			
11	Is the #21 drill 2.75 deep with a 10-32 thread .50 deep?			
12	**{Point}** 3.00 ± .015			
13	{Point} .150 Diameter ± .005			
14	{Point} .150 Diameter .25 long ± .015			
15	{10-32 Thread} .188 Diameter ± .000 -.003			
16	{10-32 Thread} .38 Long ± .015			
17	Is .25 wide flat ground to print?			
18	Is the point polished properly?			
19	Is the handle polished properly?			
20	Does the point fit properly into the handle?			

PROJECT GRADE: _____

Small Brass Hammer

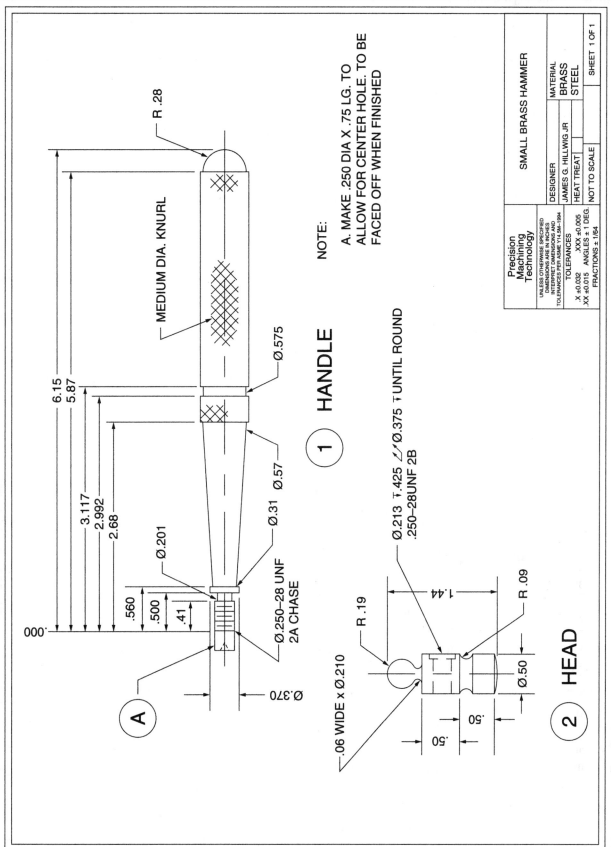

R .28

MEDIUM DIA. KNURL

Ø.575

6.15
5.87

3.117
2.992
2.68

Ø.201

Ø.31   Ø.57

.560
.500
.41

Ø.250–28 UNF
2A CHASE

.000

A

Ø.370

.06 WIDE x Ø.210

HANDLE   (1)

NOTE:

A. MAKE .250 DIA X .75 LG. TO
ALLOW FOR CENTER HOLE. TO BE
FACED OFF WHEN FINISHED

Ø.213 ⊤.425 ∠∕Ø.375 ⊤ UNTIL ROUND
.250–28UNF 2B

R .19

R .09

1.44

Ø.50

.50

.50

HEAD   (2)

	SMALL BRASS HAMMER			
Precision Machining Technology	DESIGNER		MATERIAL	
	JAMES G. HILLWIG JR		BRASS STEEL	
UNLESS OTHERWISE SPECIFIED DIMENSIONS ARE IN INCHES INTERPRET DIMENSIONS AND TOLERANCES PER ASME Y14.5M–1994	HEAT TREAT			
TOLERANCES	NOT TO SCALE			SHEET 1 OF 1
.X ±0.032 .XXX ±0.005 .XX ±0.015 ANGLES ± 1 DEG. FRACTIONS ± 1/64				

243

NIMS Skills Practiced

- Job Process Planning
- Turning Operations: Chucking
- Milling: Drilling and Tapping

Equipment

Power saw, lathe, bench grinder, milling machine

Tools Needed

Marker	6" steel rule	File card
Various files	Center drill	Knurling tool
#3 drill	1/4-28 tap	Various grits of sandpaper
Various types of turning and facing tools	Radius gages	3/8" endmill
	1/8" stamps	Hammer

Material Needed

- 5/8" diameter × 6-3/4" long CRS or aluminum (handle)
- 1/2" diameter × 1-1/2" long brass

Safety

1. Follow all shop safety rules given to you by your instructor and be sure to pass any safety tests given before using shop equipment.
2. Follow all safety rules for power equipment.
3. Be careful of sharp tools when using them.
4. Make sure the files have a handle on them.
5. Be sure all cutting tools are sharp and care is taken while handling them.
6. Be sure chuck is tightened and chuck wrench is never left in chuck.
7. Make sure stock is supported with center when necessary.

Order of Operations

HANDLE

1. Saw 5/8" Ø CRS (mild steel) 7-1/4" long.
2. Face and center drill both ends. Just face enough to get ends cleaned up.
   a. On the 1/4" – 28 end, center drill NO larger than 3/16" Ø.
3. Medium knurl 5-1/2" back.
4. On the end NOT knurled, turn .370 Ø × .850 back.
5. Turn .250 Ø × .750 back.
6. Groove .201 Ø × .090 wide as shown.
7. Thread 1/4 –28 end.
8. Face handle to length, removing center drill on radius end.
9. Turn .562 Ø back .281.
   a. The .281 radius doubled equals .562 Ø.
10. Turn .575 Ø back 2.180 from the .370 Ø shoulders.

11. Trig out angle and set compound rest to this angle.

   a. Double check angle (trig) to be sure it is correct.

12. Cut taper both directions with proper turning tools.

   a. Be careful not to cut off the .370 Ø shoulder.

13. Groove 1/8" wide ring on tapered end, deep enough to remove knurl, leaving 5/16" knurl as shown.

14. Polish the taper and the radius.

15. Stamp initials in the 1/8"-wide ring using 1/8" letters.

HEAD

1. Chuck 1/2" Ø brass bar about a foot long.

2. Face-off, then turn 3/8" Ø × 7/16" back.

3. Groove 1/16" radius to .250 Ø with radius tool.

4. Use file to form ball.

5. Groove .09 radius to .09 deep each side with radius tool.

6. Polish, then cut off head using cutoff tool to 1-1/2" long and face to 1.450 in length.

7. File convex radius on end.

   a. Watch 1.44 length.

**ONCE YOU PUT THE PART IN THE MILL, DO NOT REMOVE UNTIL STEP #11 IS COMPLETELY FINISHED!

8. On milling machine, center drill and drill with a #3 drill .430 deep.

   a. Do not break through bottom.

9. Tap 1/4 – 28 approx. 1/4" deep.

   a. Do not break through bottom.

10. Spot face with 3/8" Ø endmill, until a full 3/8" Ø circle is achieved.

11. Remove from vise and re-tap by hand with a plug tap and then with a bottom tap.

12. Saw off or grind off 1/4" – 28 threads on handle until the shoulder of handle fits brass head tightly at the spot-face.

13. Fill out evaluation sheet and turn part and evaluation in to the instructor for final evaluation and grade.

Student Name: _____ Date Submitted: _____

Class: _____ Total Hours on Job: _____

	Print Dimensions AND Tolerances	Student's Inspection	Instructor's Inspection	Instructor Comments
1	Is the part free of burrs and does it meet finish requirements?			
2	Is the part properly identified?			
3	**{Handle}** 6.15 ± .015			
4	1/4 –28 Thread .250 diameter ± .000 -.002			
5	1/4 –28 Thread .375 length ± .015			
6	Neck .201 diameter ± .005			
7	Neck .090 wide ± .005			
8	Shoulder .370 diameter ± .005			
9	Shoulder .06 width ± .015			
10	Taper 2.12 length ± .015			
11	Taper .31 diameter ± .015			
12	Taper .575 diameter ± .005			
13	Front Groove in Knurls .125 wide ± .005			
14	Front Groove in Knurls .312 back ± .005			
15	Front Groove in Knurls .575 diameter ± .005			
16	.280 radius ± .005			
17	.281 radius diameter ± .010			
18	**{Head}** 1.44 length ± .015			
19	.50 diameter ± .015			
20	.250 diameter ± .005			
21	.19 radius ± .015			

	Print Dimensions AND Tolerances	Student's Inspection	Instructor's Inspection	Instructor Comments
22	.06 radius ± .015			
23	.09 radius ± .015			
24	.50 length in center ± .015			
25	.375 Spot-face ± .005			
26	#3 Drill .425 deep ± .005			.

PROJECT GRADE: _____

Plumb Bob

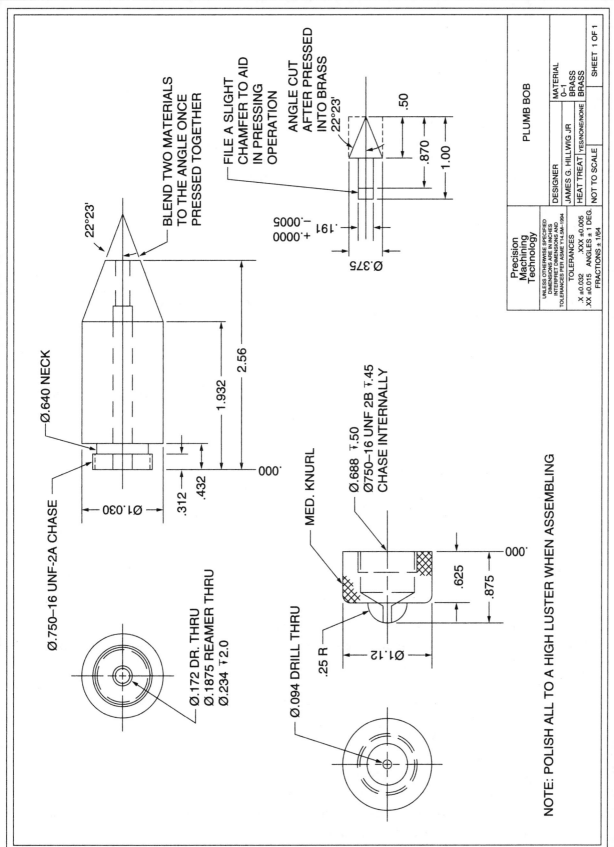

NIMS Skills Practiced

- Job Process Planning
- Turning Operations: Chucking

Equipment

Power saw, lathe with three-jaw and four-jaw chuck, bench grinder

Tools Needed

Marker	6" steel rule	File card
Various files	Center drill	Knurling tool
.625, 11/64, 3/32 drills	.1875 reamer	Indicator w/magnetic base
Various grits of sandpaper	Radius gages	Thread triangles
Various types of turning, facing, grooving, forming, and threading tools		

Material Needed

- 1-1/8" diameter × 3-5/8" long brass; 3/8" × 1" long drill rod (This needs to be heat treated 0-1; can be done with torches.)

Safety

1. Follow all shop safety rules given to you by your instructor and be sure to pass any safety tests given before using shop equipment.
2. Follow all safety rules for power equipment.
3. Be careful of sharp tools when using them.
4. Make sure the files have a handle on them.
5. Be sure all cutting tools are sharp and care is taken while handling them.
6. Be sure chuck is tightened and chuck wrench is never left in chuck.
7. Make sure stock is supported with center when necessary.

Order of Operations

BODY

1. Look over blueprint to determine tolerances, tools, and material needed.
2. Gather tools needed, figure out RPMs for required stock diameters and drills used, and write down their RPMs on sheet.
3. Measure and cut stock sizes needed.
4. Starting with the three-jaw chuck, take the 1-1/8" brass, chuck with 2" out of chuck, and face just enough off to clean up end and center drill. (Body)
5. Turn the 1.03 diameter back 1.875.
6. Turn the .750 diameter back .31.
7. Cut the .125 groove to .640 diameter .31 back.
8. Chase 3/4-16 thread (class 3A fit—check with triangles).
9. Turn end for end and hold in a four-jaw chuck (total indicator reading, or TIR, of .001 on the 1.032 diameter).
10. Center drill and knurl .875 from end on 1-1/8 diameter. (Cap)

11. Recheck TIR .001 and re-center drill if it was out.
12. Drill .562 deep from point of a 5/8 drill.
13. Now bore to .687 diameter and cut a .125 × .760 relief 1/2 back.
14. Chase a 3/4-16 internal thread.
15. Saw off cap .937 long.
16. Face part to 2.56 long. (Body)
17. Trig out angle and turn taper on the end; be careful not to go below .400 in diameter on small end.
18. Center drill and drill 11/64 through.
19. Ream .1875; there should be about a 1/32 chamfer left on hole from center drill.
20. Turn end for end in four-jaw chuck, TIR .001 on 1.03 diameter.
21. Screw on cap, face cap to clean up, center drill, and drill 3/32 through.
22. Turn .500 diameter back to allow .625 of knurls to remain.
23. Face .500 diameter to .250 long and form the .250 radius.

POINT

1. Use 3/8"-diameter tool steel.
2. Face and turn end for end and face to length.
3. Turn .191 diameter × 1/2 long.
4. File .125 long lead taper and then press into brass.
5. Hold in four-jaw chuck (TIR .001 on 1.03 diameter) and turn taper to point; this should include the brass by blending the angle seamlessly together.
6. Polish the tool steel to remove any machining marks.
7. Heat treat tip.
8. Polish all over.

Student Name: _____ Date Submitted: _____

Class: _____ Total Hours on Job: _____

		Print Dimensions AND Tolerances	Student's Inspection	Instructor's Inspection	Instructor Comments
1		Is the part free of burrs and does it meet finish requirements?			
2		Is the part properly identified?			
3		**{Part 1}** Length w/tip 3.06 ± .015			
4		Length .75 ± .015			
5		Length 1.50 ± .015			
6		Length .31 ± .015			
7		Diameter 1.03 ± .015			
8		Diameter 3/4 - 16 Thread .750 ±.000 −.005			
9		Neck .12 Length ± .015			
10		Neck Diameter .640 ± .005			
11		Through Hole 11/64 ± .015			
12		**{Part 2}** Length .875 ± .005			
13		Length .625 ± .005			
14		Depth .50 ± .005			
15		Diameter 1.12 ± .015			
16		Radius .25 R ± .005			
17		Reamed Hole .1875 ± .001			
18		Total Length 3.935 ± .015			
19		Knurl to points			
20		Angles blended on tip and body			

PROJECT GRADE: _____

Threading Project

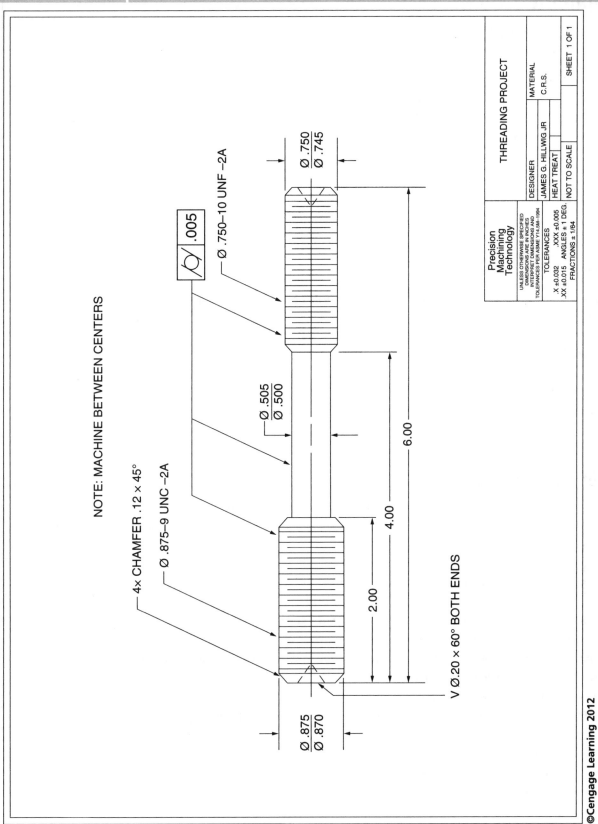

NOTE: MACHINE BETWEEN CENTERS

Ø .750–10 UNF –2A

Ø .750 / Ø .745

Ø .505 / Ø .500

4× CHAMFER .12 × 45°

Ø .875–9 UNC –2A

∅ .005

V Ø.20 × 60° BOTH ENDS

Ø .875 / Ø .870

6.00

4.00

2.00

Precision Machining Technology

THREADING PROJECT

UNLESS OTHERWISE SPECIFIED
DIMENSIONS ARE IN INCHES
INTERPRET DIMENSIONS AND
TOLERANCES PER ASME Y14.5M–1994

TOLERANCES
.X ±0.032   .XXX ±0.005
.XX ±0.015   ANGLES ± 1 DEG.
FRACTIONS ± 1/64

DESIGNER
JAMES G. HILLWIG JR

MATERIAL
C.R.S.

HEAT TREAT

NOT TO SCALE

SHEET 1 OF 1

NIMS Skills Practiced

- Job Process Planning
- Turning Operations: Chucking

Equipment

Power saw, lathe, bench grinder

Tools Needed

6" steel rule
Center drill
Various types of turning,
   facing, and threading tools

File card
Thread pitch gage
Layout dye

Various files
Go/no-go gages
Center gage

Material Needed

- Mild steel 1" × 6-1/8"

Safety

1. Follow all shop safety rules given to you by your instructor and be sure to pass any safety tests given before using shop equipment.
2. Follow all safety rules for power equipment.
3. Be careful of sharp tools when using them.
4. Make sure the files have a handle on them.
5. Be sure all cutting tools are sharp and care is taken while handling them.
6. Be sure chuck is tightened and chuck wrench is never left in chuck.
7. Make sure stock is supported with center when necessary.

Order of Operations

1. Look over blueprint to determine tolerances, tools, and material needed.
2. Gather tools needed, figure out RPMs for required stock diameters and drills used, and write down their RPMs on sheet.
3. Measure and cut stock sizes needed.
4. Place stock in chuck, face, and center drill one end.
5. Turn stock in chuck, face stock to length, and center drill.
6. Extend stock 4-1/2" from chuck face and support center in tailstock.
7. Turn .875 diameter 4" long from end.
8. Turn part in chuck and extend stock 4-1/2" from chuck face and support center in tailstock.
9. Turn .750 diameter 4" long from end.
10. Now mark the part 2-1/8" from end in tailstock.
11. Begin cutting the .500 diameter in the center without going past the 4" length you just turned and without removing the 2-1/8" mark you just made.
12. Turn part in chuck and extend stock 4-1/2" from chuck face and support center in tailstock.
13. Complete the .500 diameter 4" from end supported in the tailstock.
14. Polish the .500 diameter.

15. With your instructor's help, cut the 45° chamfers on each of the threaded sections as shown.
16. Place a lathe center in the chuck and re-cut the 60° taper on it so it runs true with the machine.
17. Place a lathe dog on one end of your part and support the other with the tailstock.
18. Paint the end you are going to thread with layout dye.
19. Set up your threading tool and machine for the proper threads per inch and touch off the part to set zeros.
20. Taking a light cut, engage the machine to thread; this should leave a light impression on the part you can check with the thread pitch gages.
21. If your thread setup is correct, finish your threads.
22. Check them along the way with thread micrometers or thread triangles.
23. Once to depth, check with go/no-go gage.
24. If gage does not fit, follow instructor's directions to pick up thread and correct the size.
25. Now turn part end for end, placing lathe dog on end you just threaded—place a penny or something over your threads where the screw touches so you do not damage the threads you just made, and place the other end in the tailstock.
26. Repeat steps 18 through 24.
27. Inspect part and turn in.

Student Name: _____ Date Submitted: _____

Class: _____ Total Hours on Job: _____

	Print Dimensions AND Tolerances	Student's Inspection	Instructor's Inspection	Instructor Comments
1	Is the part free of burrs and does it meet finish requirements?			
2	Is the part properly identified?			
3	Ø .875 ± .000 − .005			
4	9 Pitch (Threads per Inch)			
5	Ø .505 / .500			
6	Ø.750 ± .000 − .005			
7	6.0 Length ± .015			
8	10 Pitch (Threads per Inch)			
9	{7/8 Dimension} 2.0 ± .015			
10	{.50 Dimension} 2.0 ± .015			
11	{3/4 Dimension} 2.0 ± .015			
12	.12 X 45° Chamfer ends and middle			
13				
14				
15				

PROJECT GRADE: _____

Jack Screws

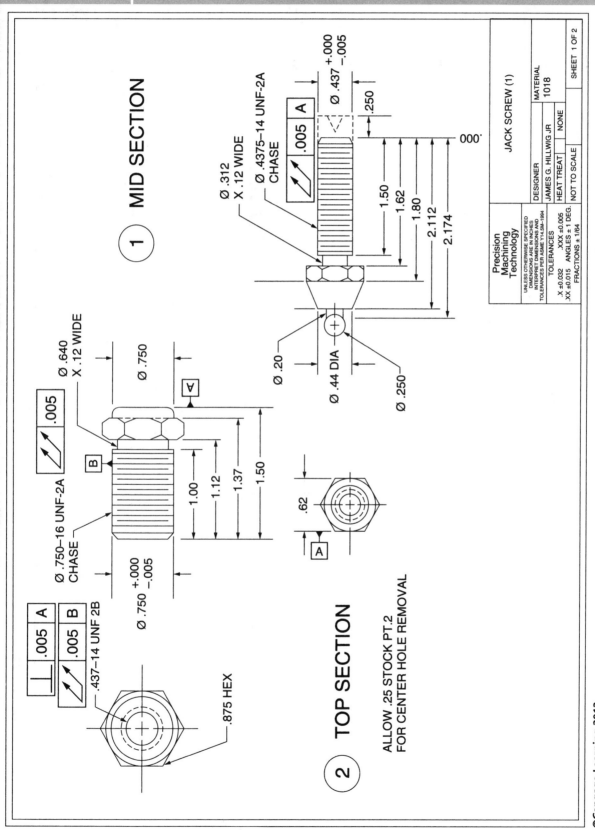

MID SECTION

1

TOP SECTION

2

ALLOW .25 STOCK PT.2
FOR CENTER HOLE REMOVAL

Ø .437 +.000 / −.005

.250

.005 | A

1.50
1.62
1.80
2.112
2.174

.000

Ø .312
X .12 WIDE

Ø .4375–14 UNF–2A
CHASE

Ø .640
X .12 WIDE

.005

B

Ø .750–16 UNF–2A
CHASE

Ø .750 +.000 / −.005

Ø .750

A

1.00
1.12
1.37
1.50

.62

A

Ø .20

Ø .44 DIA

Ø .250

| | .005 | A |
| .005 | B |

.437–14 UNF 2B

.875 HEX

	Precision Machining Technology	JACK SCREW (1)	
UNLESS OTHERWISE SPECIFIED DIMENSIONS ARE IN INCHES INTERPRET DIMENSIONS AND TOLERANCES PER ASME Y14.5M–1994	DESIGNER	MATERIAL
	JAMES G. HILLWIG JR	1018
TOLERANCES X ±0.032 .XXX ±0.005 .XX ±0.015 ANGLES ± 1 DEG. FRACTIONS ± 1/64	HEAT TREAT	NONE
	NOT TO SCALE	SHEET 1 OF 2

257

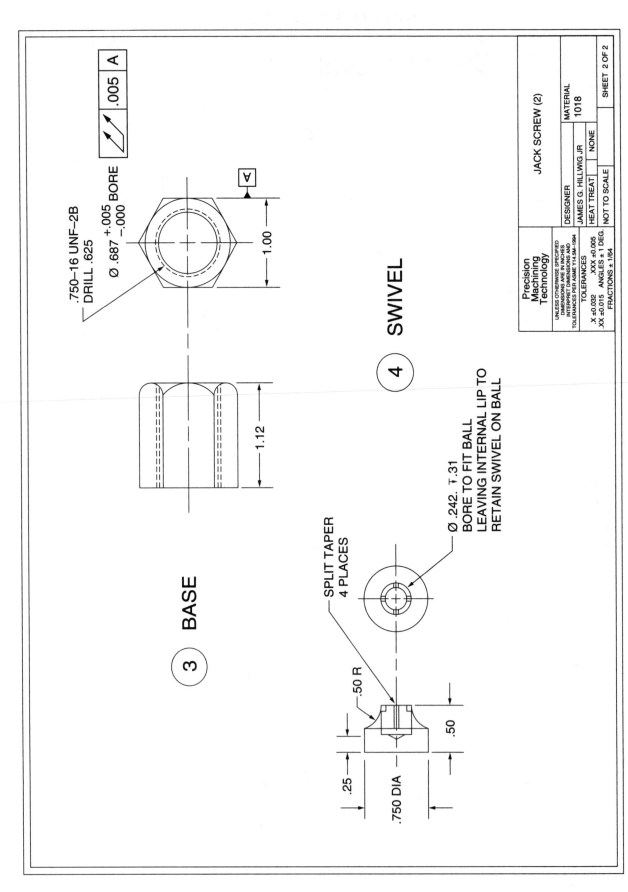

.750–16 UNF–2B
DRILL .625

Ø .687 +.005 BORE
       –.000

⟋ | .005 | A

1.00

A

1.12

BASE

(3)

SWIVEL

(4)

SPLIT TAPER
4 PLACES

Ø .242. ⊤ .31
BORE TO FIT BALL
LEAVING INTERNAL LIP TO
RETAIN SWIVEL ON BALL

.50 R

.25

.750 DIA

.50

Precision
Machining
Technology

JACK SCREW (2)

UNLESS OTHERWISE SPECIFIED
DIMENSIONS ARE IN INCHES
INTERPRET DIMENSIONS AND
TOLERANCES PER ASME Y14.5M–1994

TOLERANCES
.X ±0.032   .XXX ±0.005
.XX ±0.015   ANGLES ± 1 DEG.
FRACTIONS ± 1/64

DESIGNER
JAMES G. HILLWIG JR

MATERIAL
1018

HEAT TREAT   NONE

NOT TO SCALE

SHEET 2 OF 2

NIMS Skills Practiced

- Job Process Planning
- Turning Operations: Chucking
- Turning Operations: Between Centers

Equipment

Power saw, lathe, bench vise, bench grinder

Tools Needed

Marker	6" steel rule	File card
Files	Center drill	Various drills
Radius gages	Thread triangles	Various taps
Hacksaw	Lathe center	Lathe dog
Various types of turning, facing, grooving, forming, boring, and threading tools	Large angle plate	Shim stock
	Half nuts	Compound clamps

Material Needed

- HEX stock (CRS):
  - 1 @ 5/8" × 5-3/8" long (Part #2 Top Section)
  - 1 @ 7/8" × 3-3/8" long (Part #1 Mid Section)
  - 2 @ 1" × 1-1/8" long (Part #3 Base)
- ROUND Stock (CRS):
  - 1 @ 3/4" × 2-1/4" long (Part #4 Swivel)

Safety

1. Follow all shop safety rules given to you by your instructor and be sure to pass any safety tests given before using shop equipment.
2. Follow all safety rules for power equipment.
3. Be careful of sharp tools when using them.
4. Make sure the files have a handle on them.
5. Be sure all cutting tools are sharp and care is taken while handling them.
6. Be sure chuck is tightened and chuck wrench is never left in chuck.
7. Make sure stock is supported with center when necessary.

Order of Operations

YOU ARE MAKING TWO SETS OF JACK SCREWS

1. Look over blueprint to determine tolerances, tools, and material needed.
2. Gather tools needed, figure out RPMs for required stock diameters and drills used, and write down their RPMs on sheet.
3. Measure and cut stock sizes needed.
4. Parts #1 and #2 are made by double ending.
5. Do similar operations on parts #1 and #2 together.

6. Face and center drill ends of both pieces.

7. Part #1: between centers turn .750 diameter × 1.12 long on each end.

8. Part #2: between centers turn .437 diameter × 1.75 long on each end.

9. Part #1: hold in chuck, bring live center into centered-drilled hole, and cut .640 diameter × .12 wide neck 1.12 back from end. Flip end for end and repeat.

10. Part #2: hold in chuck, bring live center into centered-drilled hole, and cut .312 diameter × .12 wide neck 1.75 back from end. Flip end for end and repeat.

11. Part #1: between centers, chase 3/4-16 thread, class 3A fit, on each end. (Use thread triangles to check size; use half nuts to protect threads where the lathe dog touches.)

12. Part #2: chase 7/16-14 thread on both ends same as part #1.

13. Saw both parts #1 and #2 in half.

14. Part #3: mark one flat on each part with dye or a marker and always keep that flat on jaw #1. It is important to do this and use the same machine or the part will not run true enough.

15. Face both ends of both pieces and take to size 1.12 long.

16. Center drill, pilot drill, and then drill 5/8" hole through. Do both pieces.

17. Bore both pieces to .687.

18. Chase 3/4-16 internal threads to fit part #1 in both pieces.

19. Crown one end of each #3 part as shown on print.

20. Screw part #1 into part #3; make sure marked flat is lined up with jaw #1.

21. Face part #1 to length (.37" past the crowned end of part #3 or unscrew and check that it is 1.50 long).

22. Repeat on the other parts #1 and #3.

23. Turn .750 diameter by .12 long and file the 1/16" radius on both #1 pieces.

24. With marked flat on jaw #1, center drill, drill, and tap both #1 pieces.

25. Screw part #2 into part #1, which is screwed into part #3, keeping the marked flat on jaw #1.

26. Face the 5/8 hex to .812 long from part #1 and turn 1/4 diameter to .260 diameter by .312 long. Talk to your instructor about power tapping on the lathe.

27. Form ball and neck, then polish to .250 diameter.

28. Trig out angle to find the taper—same formula as used for Reversible Pocket Scribe and Plumb Bob.

29. Machine angle to 5/16 long on both pieces. Do NOT cut off corners of the hex.

30. Remove from chuck; make sure the pieces are screwed together tight.

31. Turn end for end and return to chuck with the marked flat on jaw #1.

32. Now face the bottom to a uniform appearance and the proper length; the center-drilled hole should be gone in part #2.

33. Chuck 3/4 stock in lathe; face and center drill (swivel).

34. Drill approximately .240 diameter 5/16 deep, then bore to .255, leaving about a .03 to .05 lip at the opening of the hole; this will help keep the swivel on the ball.

35. Mark 3/8 diameter on face and mark back 1/4.

36. Form the radius with a radius tool, file, polish, and check with a radius gage.

37. Turn end for end and repeat process.

38. Cut to length using a cutoff tool and deburr.

39. Hacksaw splits while holding in vise.

40. Press swivel on ball.

	Print Dimensions AND Tolerances	Student's Inspection	Instructor's Inspection	Instructor Comments
1	Is the part free of burrs and does it meet finish requirements?			
2	Is the part properly identified?			
3	**BASE Part #3** 1.12 ± .015			
4	1.0 ± .015 2.0			
5	.687 ± .005/−.000			
6	.750-16 THD Drill .625			
7	**SWIVEL** Ø.750 ± .005			
8	.50R ± .015			
9	.25 ± .015			
10	Ø.25 Drill .31 DP			
11	**MID SECTION Part #1** .62 ± .015			
12	Ø.44 ± .015			
13	Ø25 ± .015			
14	.062 ± .005			
15	.18 ± .015			
16	.312 ± .005			
17	Ø.312 × .12			
18	1.75 ± .015			
19	Ø.437 ± .005			
20	.44−14 THD			

(Continued)

	Print Dimensions AND Tolerances	Student's Inspection	Instructor's Inspection	Instructor Comments
21	**TOP SECTION Part #2** .875 HEX ± .005			
22	Ø.437−14 TAP			
23	.750-16 THD. CHASE			
24	.062 ± .005			
25	Ø.750 ±.000/−.005			
26	1.12 ± .015			
27	1.50 ± .015			
28	.12 NECK ± .015			
29	.25 ± .015			
30	.12 ± .015			
31	Ø.750 ± .015			
32	.37 ± .015			

PROJECT GRADE: _____

Flat Scribe

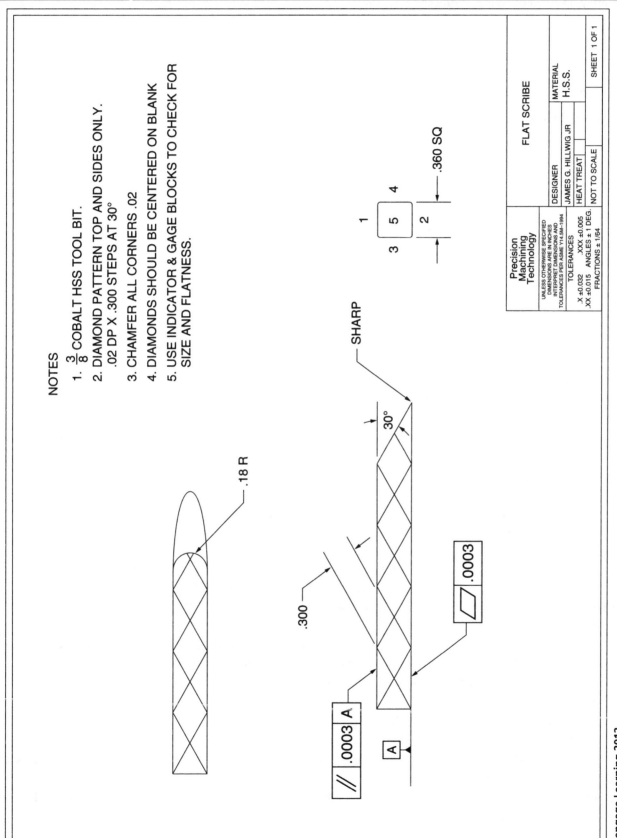

NOTES

1. $\frac{3}{8}$ COBALT HSS TOOL BIT.

2. DIAMOND PATTERN TOP AND SIDES ONLY.
   .02 DP X .300 STEPS AT 30°

3. CHAMFER ALL CORNERS .02

4. DIAMONDS SHOULD BE CENTERED ON BLANK

5. USE INDICATOR & GAGE BLOCKS TO CHECK FOR
   SIZE AND FLATNESS.

.18 R

SHARP

30°

.300

.0003

.0003 A

A

.360 SQ

3
5
1
4
2

Precision Machining Technology		FLAT SCRIBE	
UNLESS OTHERWISE SPECIFIED DIMENSIONS ARE IN INCHES INTERPRET DIMENSIONS AND TOLERANCES PER ASME Y14.5M-1994	DESIGNER	MATERIAL	
	JAMES G. HILLWIG JR	H.S.S.	
TOLERANCES .X ±0.032 .XXX ±0.005 .XX ±0.015 ANGLES ± 1 DEG. FRACTIONS ± 1/64	HEAT TREAT		
	NOT TO SCALE	SHEET 1 OF 1	

NIMS Skills Practiced

- Job Process Planning
- Surface Grinding and Safety
- Squaring up a Part on Surface Grinder

Equipment

Surface grinder

Tools Needed

Diamond dresser	Grinder vise parallels	Grinding wheel
Grinder vise	Cutoff wheel	Radius dresser
Demagnetizer	Indicator and base	Gage blocks
Surface plate	Comparator	Chuck stone

Material Needed

- 3/8" $\times$ 3/8" HSS square tool bit

Safety

1. Follow all shop safety rules given to you by your instructor and be sure to pass any safety tests given before using shop equipment.
2. Follow all safety rules for power equipment.
3. Be careful of sharp tools when using them.
4. Ring grinding wheels before using them.
5. Make sure magnetic chuck is on when surface grinding.

Order of Operations

1. Look over blueprint to determine tolerances, tools, and material needed.
2. Measure the tool bit blank to be sure it is 3/8" $\times$ 3/8".
3. Deburr all over.

**MAKE SURE MAGNETIC CHUCK IS ON WHEN SURFACE GRINDING
AND FOLLOW ALL SAFETY RULES!**

4. Check the magnetic chuck for raised burrs, and stone if necessary.
5. Mount and dress a 46-H grinding wheel.
6. Place the tool bit on the magnetic chuck at about a 40° angle to the back rail (to cover more magnetic strips), and place square blocking on both sides of the tool bit to keep it in place.
7. Be cautious of the tool bit lifting from chuck while grinding. (It will make a shearing noise if it begins to lift.)
8. Turn the magnetic chuck on.
9. Grind surface #1 until it cleans up completely.
10. Deburr the edges of side #1. BE CAREFUL: the burrs are very sharp.
11. Place side #1 on the chuck so you can grind side #2. Follow the same setup as for side #1. You can take it to size.
12. Deburr and check for flatness and size.
13. Place tool bit in grinder vise on parallels with ground sides against jaws. Leave one edge sticking out of side of vise so that you can turn vise on its side.

14. Grind top of part (side #3) to clean up and turn vise on side (side #5) without removing or disturbing part while in vise just to clean it up.

15. Remove from vise, deburr all edges, and place side #3 on magnetic chuck with square blocking on each side and grind side to size.

16. Deburr and check for flatness and size.

17. Place tool bit in grinder vise on a 30° flat angle and grind the last edge to a point.

18. Be careful not to create excessive heat while grinding angle; do not let the part change colors on you.

19. Remove the vise from the chuck. Using the radius dresser, with your instructor's help, dress a .18 radius on the wheel.

20. Place grinder vise back on chuck against the back rail and locate the center of the tool bit with the center of the radius.

21. Grind tool bit until you get a complete radius on the 30° angle.

22. Be careful not to create excessive heat while grinding radius; do not let the part change colors on you.

23. Remove vise from chuck, remove tool bit from vise, and deburr.

24. Using the same 30° angle, place the angle along the back rail and the tool bit against the angle, top up. Be sure to block the other side of the tool bit.

25. Place a cutoff wheel on surface grinder.

26. Find the corner of the tool bit and move over the part .300 plus 1/2 the thickness of the cutoff wheel. Set zero.

27. Touch-off the top of the part and set your zero. Slowly feed down .020 deep while moving table right to left.

28. Raise the wheel past zero, move table in .300, and slowly feed down .020 deep while moving the table right to left.

29. Repeat this process until you reach the other end.

30. Repeat this process on the two sides, then turn the 30° angle to make the "X" complete. You should still be able to return to the same original zero for the first line.

31. As the wheel wears down, you may need to reset your downfeed to zero.

32. DO NOT grind "X" on bottom of part.

33. Grind a .020-by-45° chamfer on all edges.

34. Inspect the part to print dimensions; record on inspection sheet and turn in.

Student Name: _____ Date Submitted: _____

Class: _____ Total Hours on Job: _____

	Print Dimensions AND Tolerances	Student's Inspection	Instructor's Inspection	Instructor Comments
1	Is the part free of burrs and does it meet finish requirements?			
2	Is the part properly identified?			
3	Are sides .360 ± .005?			
4	Is the part flat within ± .0003?			
5	Is the angle 30° ± 1°?			
6	Is the radius .18 ± .015?			
7	Are the "X's" .300 ± .005 apart?			
8	Are the "X's" to .02 ± .015?			
9	Are the "X's" centered in the part?			
10	Are the chamfers on all corners .02 deep ± .015?			
11				
12				
13				
14				
15				

PROJECT GRADE: _____

Compound Clamp

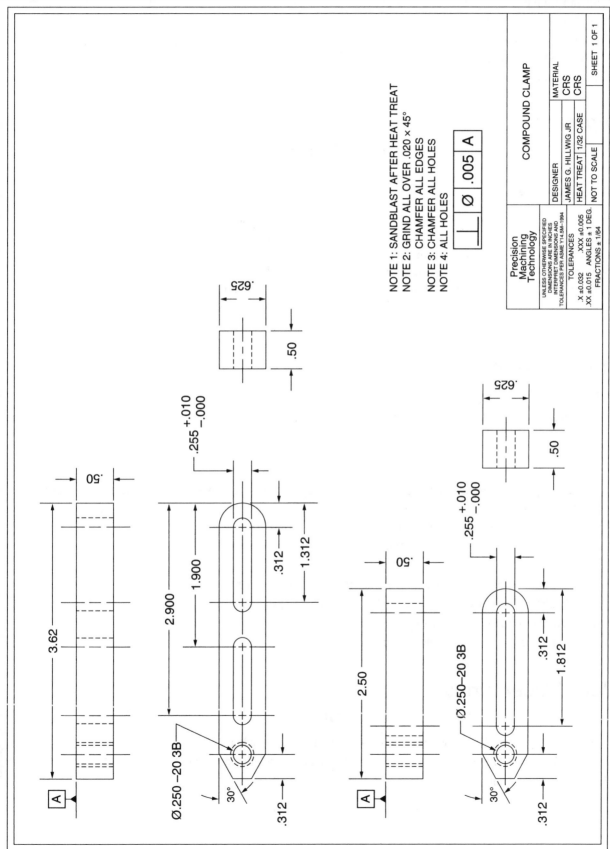

NOTE 1: SANDBLAST AFTER HEAT TREAT
NOTE 2: GRIND ALL OVER .020 × 45°
CHAMFER ALL EDGES
NOTE 3: CHAMFER ALL HOLES
NOTE 4: ALL HOLES

| ⊥ | Ø | .005 | A |

	Precision Machining Technology			COMPOUND CLAMP			
				DESIGNER		MATERIAL	
	UNLESS OTHERWISE SPECIFIED DIMENSIONS ARE IN INCHES INTERPRET DIMENSIONS AND TOLERANCES PER ASME Y14.5M–1994			JAMES G. HILLWIG JR		CRS	
	TOLERANCES			HEAT TREAT	1/32 CASE	CRS	
	.X ±0.032 .XXX ±0.005			NOT TO SCALE			SHEET 1 OF 1
	.XX ±0.015 ANGLES ± 1 DEG.						
	FRACTIONS ± 1/64						

NIMS Skills Practiced

- Job Process Planning
- Layout
- Drilling and Tapping
- Surface Grinding and Safety
- CNC Milling and Safety
- CNC Setup and Operation
- CNC Programming

Equipment

Power saw, CNC milling machine, milling machine, surface grinder

Tools Needed

Scribe	6" steel rule	Layout dye
Dead blow hammer	File	File card
Various drill bits	1/4-20 tap	Stamps
90° countersink	Various sizes of endmills	Parallels
Edge finder	Center drill	Vise stop
Machinist square	Grinding wheel	Grinder vise
Magnetic sine plate		

Material Needed

- 4 @ 3/4" square × 2-5/8" long CRS (Part #2)
- 2 @ 3/4" square × 3-3/4" long CRS (Part #1)

Safety

1. Follow all shop safety rules given to you by your instructor and be sure to pass any safety tests given before using shop equipment.
2. Follow all safety rules for power equipment.
3. Be careful of sharp tools when using them.
4. Make sure the file has a handle on it.
5. Be sure all cutting tools are sharp and care is taken while handling them.
6. Follow all CNC rules and safety guidelines for operation and programming.
7. Ring grinding wheels before using them.
8. Make sure magnetic chuck is on when surface grinding.

Order of Operations

MILL

1. Look over blueprint to determine tolerances, tools, and material needed.
2. Gather tools needed, figure out RPMs for required tools, and write down their RPMs on sheet.
3. Measure, mark, and cut stock.

4. Deburr all over.

5. Write program for cutting out parts. Use the top view for the program.

6. Load tools in correct pockets on CNC mill.

7. Place part in mill vise on parallels, leaving .650 sticking out of the top of the vise (or whatever your instructor requires), and tap down with dead blow hammer.

8. Place vise stop against edge of part, being sure it is out of the tool path.

9. Load program.

10. Edge find part and set work offsets.

11. Set tool length offsets.

12. Dry run program or single block (follow instructor's requirements).

13. Run program to cut part out.

    a. Contour outside of part.

    b. Center drill holes, tapped hole, and ends of slots.

    c. Drill tapped holes through with #7 drill.

    d. Mill slots.

    e. Repeat with remaining parts of same size.

    f. Parts of different length need a new program written.

14. The remaining material left in the vise is called the hat; you will mill that off on the manual mill and then cut your name pocket in the side, right behind the angle.

15. Set part in mill and mill to thickness.

16. Mill a name pocket centered on the 1/2" dimension just behind the angle ways .030 deep, 1/4" wide, and 3/8" long.

17. Deburr, place on anvil, and, using steel stamps, stamp your name in pocket.

This part can be heat treated if your budget permits, but it is not necessary.

SURFACE GRINDER

MAKE SURE MAGNETIC CHUCK IS ON WHEN SURFACE GRINDING!

18. Surface grind all sides square and parallel. Deburr after each side is ground.

19. Now check sizes, hole location, and slot; grind them into location and part to size. Be sure to deburr and block part when needed.

20. Using vise and magnetic sine bar, calculate gage block height, set magnetic sine bar to height, place part in grinder vise, place on magnetic sine bar, TURN ON MAGNETS, and grind angle.

21. Using a sine bar and roll dimensions, check the angle. Have your instructor help you.

22. Lay flat on grinder chuck and grind a .030-by-45° chamfer on all edges.

23. Evaluate and turn in.

Student Name: _____ Date Submitted: _____

Class: _____ Total Hours on Job: _____

	Print Dimensions AND Tolerances	Student's Inspection	Instructor's Inspection	Instructor Comments
	Part #1—Support			
1	Is the part free of burrs and does it meet grind finish requirements?			
2	Is the part properly identified?			
3	2.800 in. length ± .005			
4	1.900 in. length ± .005			
5	1.00 in. length ± .015			
6	5/16 in. radius ± .015			
7	3 5/8 in. length ± .015			
8	5/8 in. length ± .015			
9	1/2 in. length ± .015			
10	.250 width +.010 /-.000			
11	30° angle ± 1°			
12	1/4-20 3A THD. Good form?			
13	5/16 in. THD. position ± .015			
	Part #2—Clamp			
14	Is the part free of burrs and does it meet finish requirements?			
15	Is the part properly identified?			
16	2-1/2 in. length ± .015			
17	5/8 in. length ± .015			
18	1/2 in. length ± .015			
19	30° angle ± 1°			
20	1/4-20 3A THD. Good form?			

	Print Dimensions AND Tolerances	Student's Inspection	Instructor's Inspection	Instructor Comments
21	5/16 in. THD. position ± .015			
22	.250 width +. 010 /−.000			
23	1-1/2 in. length ± .015			
24	5/16 in. radius ± .015			

PROJECT GRADE: _____

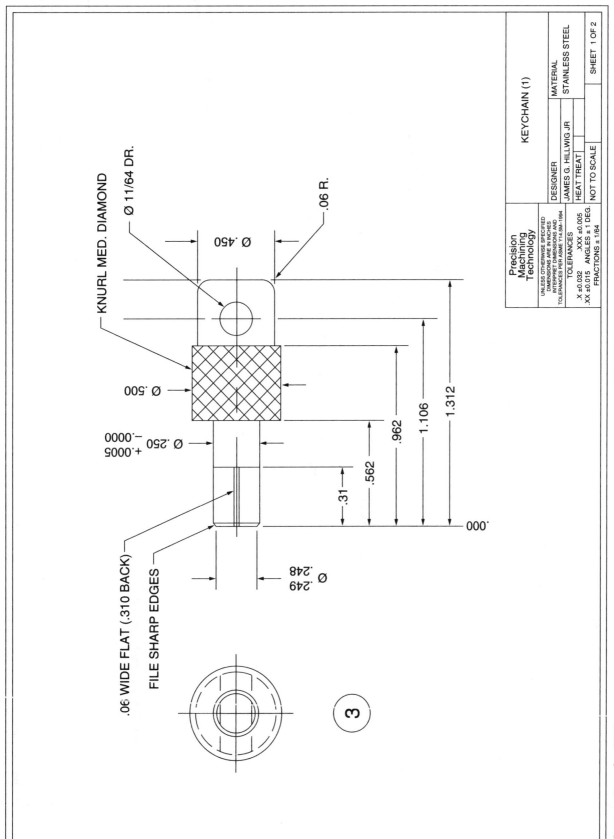

KNURL MED. DIAMOND

Ø 11/64 DR.

Ø .450

.06 R.

Ø .500

Ø .250 $^{+.0005}_{-.0000}$

.562

.31

.962

1.106

1.312

.000

Ø .249 / .248

.06 WIDE FLAT (.310 BACK)

FILE SHARP EDGES

3

KEYCHAIN (1)		
DESIGNER	MATERIAL	
JAMES G. HILLWIG JR	STAINLESS STEEL	
HEAT TREAT		
NOT TO SCALE	SHEET 1 OF 2	

Precision Machining Technology

UNLESS OTHERWISE SPECIFIED
DIMENSIONS ARE IN INCHES
INTERPRET DIMENSIONS AND
TOLERANCES PER ASME Y14.5M–1994

TOLERANCES
.X ±0.032   .XXX ±0.005
.XX ±0.015   ANGLES ± 1 DEG.
FRACTIONS ± 1/64

273

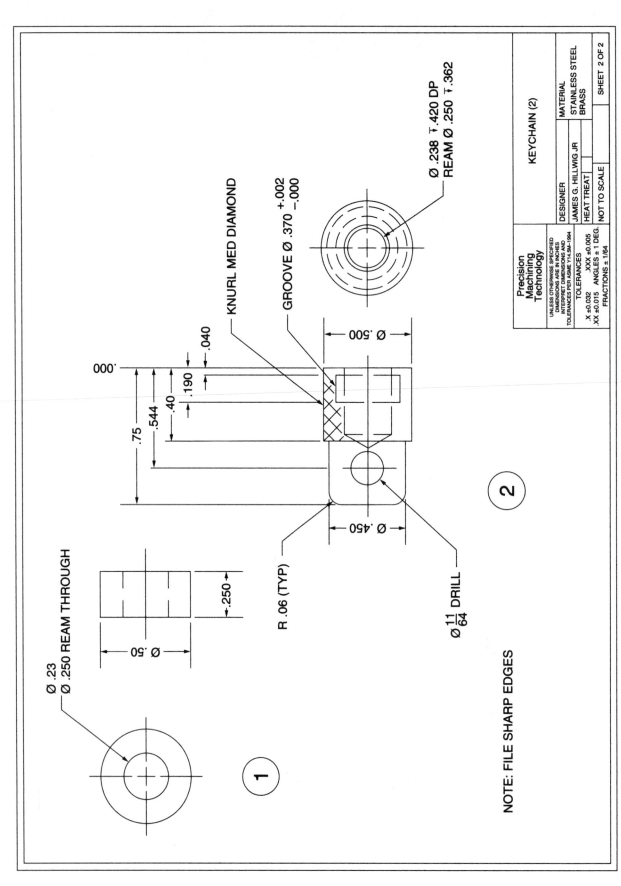

KNURL MED DIAMOND

Ø .238 ⊤ .420 DP
REAM Ø .250 ⊤ .362

GROOVE Ø .370 +.002 / −.000

Ø .500

.040

.190

.544

.40

.75

.000

Ø .450

R .06 (TYP)

Ø 11/64 DRILL

2

Ø .23
Ø .250 REAM THROUGH

Ø .50

.250

1

NOTE: FILE SHARP EDGES

Precision Machining Technology		KEYCHAIN (2)	
UNLESS OTHERWISE SPECIFIED DIMENSIONS ARE IN INCHES INTERPRET DIMENSIONS PER ASME Y14.5M-1994 TOLERANCES .X ±0.032 .XXX ±0.005 .XX ±0.015 ANGLES ± 1 DEG. FRACTIONS ± 1/64	DESIGNER JAMES G. HILLWIG JR	MATERIAL STAINLESS STEEL BRASS	
	HEAT TREAT		
	NOT TO SCALE		SHEET 2 OF 2

NIMS Skills Practiced

- Job Process Planning
- Turning Operations: CNC Chucking

Equipment

Power saw, CNC lathe, mill, bench grinder

Tools Needed

Marker	6" steel rule	File card
Various files	Center drill	Knurling tool
Various drills	.260 reamer	Radius gages
Various types of turning, facing, boring, and forming tools		

Material Needed

- 1/2" diameter × .250" long brass (ask your instructor if you should use a longer piece to hold safely in chuck, and then cut it off)
- 1 @ 1/2" diameter × 1." long stainless steel
- 1 @ 1/2" diameter × 1.5." long stainless steel
- 2 @ 1/4" O-rings (insert them in the bored groove; this will hold both pieces together)

Safety

1. Follow all shop safety rules given to you by your instructor and be sure to pass any safety tests given before using shop equipment.
2. Follow all safety rules for power equipment.
3. Be careful of sharp tools when using them.
4. Make sure the files have a handle on them.
5. Be sure all cutting tools are sharp and care is taken while handling them.
6. Follow all CNC rules and safety guidelines for operation and programming.
7. Be sure chuck is tightened and chuck wrench is never left in chuck.
8. Make sure stock is supported with center when necessary.

Order of Operations

PART #1

1. Look over blueprint to determine tolerances, tools, and material needed.
2. Gather tools needed, figure out RPMs for required stock diameters and drills used, and write down their RPMs on sheet.
3. Measure and cut stock sizes needed.
4. Write programs for cutting out parts.
5. Load tools in turret on CNC lathe.
6. Place part #1 in chuck with about 1/2" sticking out.
7. Load program.

8. Set tool length offsets.

9. Dry run program or single block (follow instructor's requirements).

10. Run program to cut part out.

    a. Face part.

    b. Center drill hole.

    c. Drill and ream.

    d. Cut off part.

PART #2

1. Place part #2 in chuck with about 1" sticking out.

2. Load program for part #2.

3. Set tool length offsets.

4. Dry run program or single block (follow instructor's requirements).

5. Run program to cut part out.

    a. Face part.

    b. Knurl part.

    c. Center drill part.

    d. Drill .238-diameter hole.

    e. Ream hole to .250 diameter.

    f. Bore internal groove.

    g. Cut off part.

6. Place part back in chuck, hole first.

7. Set tool length offsets.

8. Dry run program or single block (follow instructor's requirements).

9. Run program to cut .450 diameter by .350 long with .06 radius.

PART #3

1. Place part #3 in chuck with about 1.125" sticking out.

2. Load program for part #3.

3. Set tool length offsets.

4. Dry run program or single block (follow instructor's requirements).

5. Run program to cut part out.

    a. Face part.

    b. Knurl part.

    c. Turn .250 diameter by .562 long.

    d. Turn .248 to .249 by .31 long.

    e. Chamfer while cutting off part.

6. Place part back in chuck, pin side first.

7. Set tool length offsets.

8. Dry run program or single block (follow instructor's requirements).

9. Run program to cut .450 diameter by .350 long with .06 radius.

10. Set **part #2** in mill, center drill, and drill 11/64-diameter hole.

11. Set **part #3** in mill, center drill, and drill 11/64-diameter hole.

12. Mill or file .06 wide flat .310 as shown.

Student Name: _____ Date Submitted: _____

Class: _____ Total Hours on Job: _____

	Print Dimensions AND Tolerances	Student's Inspection	Instructor's Inspection	Instructor Comments
1	Is the part free of burrs and does it meet finish requirements?			
2	Is the part properly identified?			
3	**{Part #1 Brass}** .50 Diameter ± .015			
4	(Hole Diameter) .250 ± .005			
5	(Length) .250 ± .005			
6	**{Part #2 Stainless Steel}** .500 Diameter ± .005			
7	(Length) .75 ± .015			
8	(Diameter) .450 ± .005			
9	(Radius) .06 ± .015			
10	(Length) .350 ± .005			
11	(Internal Hole Diameter) .250 ± .005			
12	(Internal Hole Length) .420 ± .005			
13	(Internal Groove Diameter) .370 ± .002 -.000			
14	(Internal Groove Width) .150 ± .005			
15	(Internal Groove Lip) .040 ± .005			
16	(Drilled Hole Diameter) 11/64 ± 1/64			
17	(Drilled Hole Location) .206 ± .005			
18	**{Part #3 Stainless Steel}** .500 Diameter ± .005			
19	(Length) 1.312 ± .005			
20	(Diameter) .450 ± .005			
21	(Radius) .06 ± .015			
22	(Length) .350 ± .005			

(Continued)

	Print Dimensions AND Tolerances	Student's Inspection	Instructor's Inspection	Instructor Comments
23	(Knurl Length) .40 ± .015			
24	(Internal Hole Length) .420 ± .005			
25	(Pin Length) .562 ± .005			
26	(Pin Diameter) .249 to .248			
27	(Pin Length with Flats) .31 ± .015			
28	(Pin Diameter) .250 ± .001 - .000			
29	(Pin Flats) .06 wide ± .015			
30	(Drilled Hole Diameter) 11/64 ± 1/64			
31	(Drilled Hole Location) .206 ± .005			

PROJECT GRADE: _____